Navigating Through
a Stipulated Freedom

Navigating Through a Stipulated Freedom

Discovering a Guiding Biblical Compass
For the Journey of Biotechnology

Paul J. Kirbas

Cloverdale Books
South Bend

Navigating Through a Stipulated Freedom: Discovering a
Guiding Biblical Compass For the Journey of Biotechnology

Paul J. Kirbas

Published by
Cloverdale Books
An Imprint of Cloverdale Corporation
South Bend, Indiana 46601

www.CloverdaleBooks.com

Library of Congress Cataloging-in-Publication Data
Kirbas, Paul J.
  Navigating through a stipulated freedom : discovering a guiding biblical
compass for the journey of biotechnology / Paul J. Kirbas.
      p. ; cm.
  Includes bibliographical references and index.
  Summary: "This book demonstrates clear biblical and theological evidence
that supports the advancements of biotechnology that so often get
restricted by religious convictions. As the core values that the Bible offers as
the guiding directives are articulated, the book develops an ethical compass
that is then applied to some of the most controversial issues of our day"--
Provided by publisher.
  ISBN-13: 978-1-929569-39-7 (pbk.)
  ISBN-10: 1-929569-39-4 (pbk.)
  1. Biotechnology--Religious aspects--Christianity. 2. Bible. I. Title.
  [DNLM: 1. Bible. 2. Biotechnology--ethics. 3. Christianity. 4. Religion and
Science. TP 248.23 K58n 2007]

TP248.23.K57 2007
174'.96606--dc22
                                    2007023041

Printed in the United States of America
on recycled paper made from 100% post-consumer waste

 # Table of Contents

Preface   vii

Acknowledgements   ix

Introduction   1

1   The Sacred and Nature   9

2   Discovering the Yellow Light   39

3   Creating the Compass   69

4   The Compass Applied to Plant
and Animal Biotechnology   97

5   The Compass Applied to Stem Cell Research
and Prenatal Engineering   117

6   The Compass Applied to Bionics,
Artificial Intelligence, and Cybernetic Immorality   141

7   Conclusion
Religion Engaging Science in a Positive Way   165

Bibliography   181

Index   191

# Preface

The purpose of this book is to design an evaluative compass, based on a clearly defined Biblical ethic, by which key issues of biotechnology can be evaluated for ethical purposes. The book begins in chapter one by establishing the argument that in the Judeo-Christian faith, the natural realm is not intrinsically sacred, but has been made sacred by God as a tool of self-revelation and for the dispensing of grace. If nature is made sacred by God, then only God has the right to establish parameters for human manipulations of nature.

The second chapter of this book builds on this idea, and identifies a pivotal Biblical location for an understanding of the parameters of God's permission for humans to manipulate nature. This understanding comes from the Decalogue, particularly the commandment concerning the proper use of God's name. The commandment, understood in the larger context that is explained in chapter two, offers humanity freedom to manipulate nature as long as God's core values are preserved. Activities that oppose those core values, as well as activities that fail to promote them, are in essence, using God's name *in vain*.

The third chapter sets out to build the actual compass that is used as an evaluative tool for biotech issues. The chapter begins by seeking to identify two concepts that are, by many views, core values of the Biblical text. These are identified as *love* and *justice*. The two terms are explained and explored according to their Biblical use. The remainder of chapter three builds a compass that is demonstrated to the reader in a series of drawn figures. Once the compass is fully presented, the book moves forward to use the compass in evaluating some major issues of biotechnology.

In chapter four, the compass is applied to the issues of plant and animal biotechnology. In chapter five, the compass is applied to stem cell research and to prenatal engineering. Chapter six addresses a group of issues related to mechanical and digital biotechnology, namely bionics, artificial intelligence, and cybernetic immortality.

As the compass is applied to all of these issues, this book actually draws the arrow of the compass in a relationship to God's core values based on the considerations of each chapter. The total book, therefore, has repeated figures of the compass, in which the arrow is pointed in certain directions. It is hoped that as the reader gets comfortable seeing how the compass could be applied, he/she will use the compass to evaluate other issues not specifically addressed in this book.

The final chapter offers tangible ways in which the religious community could, utilizing the compass, become a partner with government and science in addressing the future course of biotechnology. Before getting to those suggestions, the chapter will identify certain obstacles that would need to be addressed in order for the compass to gain significant use.

The brief conclusion at the end of chapter seven offers the hope that the best dreams of biotechnology will be realized, and that in doing so, God's core values of love and justice will grow and multiply...indeed to cover the earth...by them. In so doing, human potential could be realized, and God's stipulated freedom would be fully embraced.

# Acknowledgements

Writing this book has been a labor of love for me, and quite a journey that has included a number of special people. I am deeply grateful for all of the support, encouragement, and sharing of experiences that many have offered to me. Although there are many people that have been helpful, I would like to mention a few that have played significant roles.

I wish to acknowledge my PhD advisor John Kerr, who first sparked my interest in bioethics as I attended his Oxford University class many years ago. Since that time, he has been a wonderful mentor and friend. I wish to offer my deepest appreciation for my "A-team", a group of retired doctors, professors, and corporate executives who served as an advising and accountability group. These included Dr. Lee Greene, Bob Cutler, Dr. Bob Windom, and Dr. David Thomas. I am grateful for the counsel and advice offered by Dr. John Morgan, President of the Graduate Theological Foundation. Dr. Morgan is a great encourager, and kept me on track during the most difficult days of my PhD program. I would like to thank Dr. Charles Scouten, with whom I have had numerous email and face to face conversations concerning issues of biotechnology that have helped keep the fire burning within my heart and mind.

I am thankful for the congregations that have embraced my teaching programs, and have enabled me to build expertise and confidence as a teacher of theology and biotechnology. These include the First Presbyterian Church of Bonita Springs, Florida; Church of the Palms in Sarasota, Florida; and the First Presbyterian Church of Wheaton, IL.

Finally and most importantly, I want to thank my wife Jennifer and two sons, Jeff and Brad, for giving me the time and space to complete this book. It is the support of my family that has enabled me to achieve many things in life.

Introduction

The 21st century is destined to be an age of incredible possibilities in the realms of science and technology, specifically as these disciplines give humanity unprecedented control of the biology of life. This new level of power to manipulate nature is, in many ways, outpacing our traditional notions of theology and ethics. As this happens, the religious community will have two options. The first is to assume that anything that moves beyond our traditional understandings of proper boundaries must be inherently wrong. This view will often apply the term "playing God" to efforts that go beyond traditionally accepted boundaries, with the assumption that any human encroachment upon "God's territory" is forbidden and evil. This view places religion as a constant opponent to science and biotechnology, and fuels many of the political debates that rage in our country on issues such as cloning and stem cell research. There is, however, a second possible approach. If our traditional theologies and ethics are unable to speak relevantly to the biotech developments of the 21st century, there seems to be the urgent need to find new theological and ethical ground. This is not to suggest that the basic foundations of our faith and morality should be abandoned. Rather, one should bring new questions to those foundations, and find in them a hidden power. It is the power of long standing religious ideas to speak relevantly and meaningfully to the radically new manipulations of biology and nature. This book will seek to establish a pathway between core theological concepts and the advancing world of biotechnology.

It should be noted from the onset that the grounding of ethics related to aspects of biotechnology in the realm of Judeo-Christian theology assumes that one accepts this theology, and its supporting Biblical texts, as authoritative. This book, then, requires a religious mindset. As such, there will be a large segment of people that will not be willing to make the fundamental assumptions presented here.

Many in the scientific community, for instance, have argued for a clear separation of scientific inquiry from any religious entanglement. They do so, one must admit, for good reasons. For too long, religion has been the rain on the parade of scientific progress. One only has to recall the case of Galileo, or the Scopes trials of last century, or of the current struggle over teaching evolution and/or intelligent design theory in public school classrooms, and it is easy to see why such a wall of resistance has been built between science and religion. Some in the evolutionary biology field are using their time and talents to discredit religion as much as possible. In his recent book Darwin's Cathedral, David Sloan Wilson builds a strong case for seeing religions as nothing more than the group selection process of evolution, much like the organization observed in ants and bees. The newest book of the philosopher Daniel Dennett, entitled Breaking the Spell; *Religion as a Natural Phenomenon*, is making a similar point. Stephen Jay Gould, in his book Rocks of Ages, advocates a respectful distance between science and religion, asking that we follow the principle of NOMA, which stands for "Non-Overlapping Magisteria".[1] In other words, religion should stay in its place, which is under the steeple and not in the laboratory. Gould does recognize the need for the integration of ethics and actions, but largely sees that as an integration that happens in the heart and soul of each individual rather than a more formal interaction of religion and science. These ideas are some of the most favorable toward a religion-science interface by leaders in the fields of evolutionary biology. For a real slap in the face of religion, there are the words of Dr. Richard Dawkins of Oxford University. In an article published in the January/February 1997 edition of *The Humanist,* Dawkins began with these words:

> *It is fashionable to wax apocalyptic about the threat to humanity posed by the AIDS virus, "mad cow" disease, and many others, but I think a case can be made that faith is one of the world's great evils, comparable to the smallpox virus but harder to eradicate.* [2]

It would be too lofty a goal to aspire that one could establish a theological basis for the ethics of biotechnology that would be seen as authoritative, or even influential, by those who hold such positions. For the non-religious, the ethics that govern biotechnology would have to come from an entirely different source; no God, no scriptures, and no religion. When these traditional sources of authority are discounted, where does one find grounding for ethics and morality? In his book <u>Godless Morality</u>, Richard Holloway attempts to paint a picture of what one such non-religious morality would look like. He writes:

> What this suggests, therefore, is the responsible but exciting possibility of rethinking morality for our own day, acknowledging our situation, its confusions and insights, while also recognizing that we need order and balance in our lives. But today, perhaps for the first time, we shall struggle to achieve a morality that is self-imposed and consented to by our own reasons, though even that will not guarantee our compliance.[3]

Developing the ethics to accompany biotechnology from a self-imposed rule of consensus may be the best that a Godless morality could provide, and perhaps the only ethical argument that some would accept. There are, however, many others who do believe that there is something higher than human consensus. For many, there is a God who has expressed a will and a guide for all human endeavors. For many religiously inclined people, this expression of God's will can be found in the basic texts of the Judeo-Christian faith. This book is written for those who hold this conviction.

The problem with anchoring an ethical response to biotechnology in Judeo-Christian theology, however, is that the traditional interpretations of this faith tradition are being quickly outdated by the advances of science. Christian theology is ill prepared to recognize humanity's ability to create new forms of life, never (to our knowledge) intended by the Creator. Christian anthropology is ill prepared to understand the integration of biology and technology, in which the lines between human and machine will become increasingly blurred. Christian eschatology is ill prepared to address the bridging of neuron and digital signals, opening up the possibility of "uploading" a human mind into a kind of cyber-immortality.

As long as the Christian response to biotech possibilities clings to historic theological assumptions, we will soon discover our kinship to the religious leaders who insisted that the earth was the center of the universe, in spite of Galileo's growing evidence to the contrary[4], or to the religious leaders who even today resist the mounting evidence of natural selection because it does not perfectly align with the Genesis account of creation. To think that we can take on this partnership with science, without undergoing significant change within our own Christian identities, is severely misguided.

In order to meaningfully engage biotechnology, then, the Christian faith must undergo some significant changes in two major areas. First, we must be willing to entertain changes in our *theology*. Long held beliefs about the nature of God, the world, and ourselves will need to be challenged. Some would cringe at that suggestion, seeing our classic Christian theology as sacred in itself. We must remind ourselves, however, that the word "theology" means "the study of God". The *study of* God is not God. Theology is a human endeavor, and as such, is open to mistakes and corrections.

The second area of Christian faith that must undergo changes in order to meaningfully engage biotechnology is our *ethics*. This area is purposely placed second, because it grows out of the first. Ethics can be defined as a set of rules and standards that govern human behavior. But just as our public rules (i.e. laws) grow out of a larger philosophy (i.e. a constitution) so too, our ethics grow out of a larger philosophy, which is our theology. New takes on theology will breed new ethics in application.

If ways can be identified in which the core concepts of a religious morality can engage the advancing field of biotechnology in an enlightened and positive relationship, a new path will be found between the wisdom of the ages and the technologies of the future. In seeking to establish this path, one must first identify a basic source of religious instruction that is generally seen as establishing God's parameters of human activity. For all three major monotheistic faiths (Christianity, Islam, and Judaism), one source does seem to take primary precedence. It is known as the "Ten Commandments". Within the commandments, this book proposes, one can find a stipulated freedom offered by God. This freedom is not absolute, however. It has certain limits. It is a *stipulated* freedom. This stipulated freedom invites humanity to travel down the yet-to-be discovered road of present and future developments of biotechnology, albeit within certain parameters.

When one travels in an unknown direction, it is always helpful to have some guide available. Ideally, that would be a map, which could clearly identify the right road to travel. Such a map could show which turns to take, and which ones to avoid. It is hard to go wrong when you have a map. As we travel down the road of biotechnology, it would be nice to have such an ethical map. Unfortunately, no such map exists. The truth is that the path ahead is truly uncharted territory. If we are unable to draw a map, perhaps we could at least benefit from something akin to a compass; something that points us in the general direction of values desirable and good.

This is what is proposed in this book. First, we will take a new look at the theology of sacred space. We will look at some Biblical models for understanding the concept of sacred space, and will interpret those models in light of the frontiers of biotechnology. Once we have established our working definitions of sacred space, we will explore a Biblical precept that provides evidence that God has given humanity a freedom to operate within that space, provided that certain stipulations are followed. These stipulations will be identified and researched, with the goal of creating the compass by which we could evaluate the many ethical questions raised by the possibilities of biotechnology.

Having articulated a theology of sacred space, discovered a Biblical precept for our human freedom within that space, and highlighted the stipulations that God has placed on that freedom, we will move forward with compass in hand to address some of the subjects of biotechnology that are posing the most difficult ethical challenges. As metaphorical symbols to group and organize the subjects of each chapter, a particular member of the ancient Greek Pantheon will serve as an introductory metaphor for each issue addressed. In doing this, we must realize from the start that using metaphors from ancient sources to address modern issues of biotechnology has its limits. This was well articulated by Rebecca Proefrock in her essay *The Perils of Metaphor in the Science and Religion Dialogue*, in which she critically reviewed the work of Byron Sherwin's <u>Golems Among Us: How a Jewish Legend Can Help Navigate the Biotech Century.</u> Proefrock writes:

> This book will argue that the use of metaphors-any metaphor-is plagued with inherent authority issues which the very reach for the metaphor can obscure. Moreover, by discarding items from a given metaphor, that same metaphor is removed from its *Sitz in Laben* and threatens to become useless.[5]

The use of Greek mythology in this book might receive the same criticism from Proefrock. The purpose here, however, is simply illustrative. Without relying too heavily on the total mythology represented by these figures, they are offered as simple images to collect and organize our thinking. The metaphors should not be pushed beyond their illustrative functions.

Pegasus will begin the Pantheon parade. Most people would immediately recognize this fascinating horse, which has a major difference. Although he is clearly of the *Equus Caballas* species, he has one feature that would seem to fit him into the *Aves* species. He is a horse with wings. Pegasus will introduce the first line of biological manipulation. In this chapter, the compass will be applied to the emerging world of plant and animal bioengineering.

The Greek figures *Hermes* and *Hercules* will be the mascots for our second set of issues. In this chapter, we will apply our compass to two major issues of human biotechnology. Hermes, the herald of the Olympian gods, was the most amazing newborn the world had ever seen. On the very first day of his birth, Hermes ran to a far off field in Thessaly and stole a herd of cows, which he drove all the way back to Greece. Even though it was obviously an already busy night for the newborn, he still had time to invent the lyre. If Hermes had such abilities and powers on the first day of life, it seems safe to say that as a fetus, even an embryo, he was storing some amazing powers and potential. The powers and potential of that embryo invite us to consider the power and potential of all embryos, particularly due to the stem cells they contain. In the first part of this chapter, we will address the theological and ethical issues related to stem cell research.

A human by maternal blood, Hercules was also endowed with the genes of the gods. This gave him amazing powers of strength and intellect. His mere human counterparts had every right to be jealous, for after all, Hercules had an unfair advantage. He had designer genes. In the second part of this chapter, the topic to be discussed is the genetic modification potential of human progeny. Using the compass, the theological and ethical implications of prenatal engineering will be explored.

The final figures of Greek mythology to guide us will be *Hephaestus* and *Hades*. Hephaestus was an amazing craftsman, and was able to invent gear that became integrally connected with the recipient. He was the one who equipped Zeus with thunderbolts, arrows for Eros, the chariot of Helios, and the invincible armor of

Achilles. Before long, Hephaestus was becoming so skilled at his craft that it was hard to distinguish between machine and human, especially when he built a woman named Pandora. Hephaestus will serve as the introduction to the issues of bionics and artificial intelligence. As machine and biology fuse together, we will be faced with tremendous challenges to our theological assumptions of what it means to be a human. And if all that remains in defining a human is our intelligence, how do we classify an artificial intelligence. Is it human, or machine? A half century ago, Alan Turing presented a test to evaluate the thought process of machines.[6] In the next 50 years, many more questions will emerge. These questions, along with important issues of ethics, will be the challenge in this section of the book.

In the final application of this chapter, *Hades* will lead the way. Most are familiar with Hades as the god of the underworld. He invites us to think about issues of mortality, and immortality. In this section, we will discover some of the most radical ideas of biotechnology, namely mental uploads and cyber-immortality. In this strange world of Hades, some very definite issues emerge in our theology and our ethics.

This, then, will bring us to the final chapter of this book. In this final section, this book will seek some concrete ways in which the Christian faith can play a role in determining the direction of our journey of possibility that biotechnology presents. The Greek mythology figure invoked here will be that of *Prometheus*, whose name implies "forethought".

As can be surmised by these topics, this book will address not only the most plausible biotechnology *probabilities*, but also some of the most radical *possibilities*. Some of these ideas will seem preposterous to the reader, who will wonder why one would waste time and energy addressing them. Perhaps the reason lies in a strong belief in human capability. When it comes to the concept of human potential, one can be guided by one of the early stories of the Hebrew Scriptures. In the book of Genesis, there is a story of the human community attempting to build a tower to reach the realm of heaven. It was an early human attempt to invade a sacred space. In Genesis 11, we read:

> The LORD came down to see the city and the tower, which mortals had built. [6]And the LORD said, 'Look, they are one people, and they have all one language; and this is only the

beginning of what they will do; nothing that they propose to do will now be impossible for them. [7]'Come, let us go down, and confuse their language there, so that they will not understand one another's speech.' [8]So the LORD scattered them abroad from there over the face of all the earth, and they left off building the city.[7]

In this story, God knew the human potential. God knew that nothing humans propose, nothing they dream, will be impossible for them. God, therefore, confused their language, to keep them from invading sacred space.

In the evolving sciences of biotechnology, we're back on track with the tower, and we're headed for that sacred space. What exactly is sacred space? How can one interpret this in terms of the manipulations of nature present in the work of biotechnology? This is the subject that will begin this journey as we now turn to chapter one.

Introduction Notes

[1] Gould, Stephen Jay. <u>Rocks of Ages</u>. *Science and Religion in the Fullness of Life.*, p.5

[2] This quote was taken from the website http://www.TheHumanist.org.

[3] Holloway, Richard. <u>Godless Morality</u>: *Keeping Religion out of Ethics*, p. 32

[4] The debate between Galileo and the church was not as simple as the common understanding would suggest. It seems that Galileo's arrogant portrayal of Pope Urban VIII's views as simplistic was a large contributing factor to his problems with the church. For a full exploration of this issue, see <u>Galileo's Mistake</u>. *A New Look at the Epic Confrontation between Galileo and the Church*, by Wade Rowland.

[5] Proefrock, Rebecca, In a paper entitled *The Perils of Metaphor in the Science and Religion Dialogue*, presented to the *Chicago Group* Tuesday, March 6, 2007.

[6] The Turing Test was presented by Alan Turing in a 1950 paper entitled *Computing Machinery and Intelligence*. The test basically suggested that if a judge is unable to differentiate between a human and machine participant in a conversation, the machine passes the test and can be labeled as an "intelligent machine."

[7] Unless otherwise noted, all Biblical references in this book are from the New International Version

The Sacred and Nature

Introduction

The mountain was mysteriously smoky. Down below, a shepherd was tending the sheep. His name was Moses. The mysterious smoke on the mountain seemed intriguing to him, so Moses decided to investigate the scene. As he traveled up the mountain, he suddenly came upon a startling sight. A bush seemed to be engulfed in flames, but oddly enough, was not being consumed by the fire. As Moses looked upon this sight, he heard a voice call out from the center of the burning bush. It was the voice of God, who said:

> *Do not come near this place. Take your sandals off your feet,*
> *For the place where you stand is holy ground.*[8]

From this well known story of the Hebrew Scriptures, two central thoughts emerge. They are the very concepts that are at the heart of the ethical debates that rage over the issues of biotechnology. The first concept is that of *sacredness*. The reason Moses was told to take off his sandals was because he was approaching the sacred. Approach it he may, but he is forbidden to enter it. That would be easy enough to follow if the sacred was *spiritual*, that is, other than physical. That leads to the second concept. In Moses' case (as in the modern day), the sacred is not merely spiritual. God said, "The *place* where you stand is holy *ground.*" The forbidden territory of sacredness is a definable, measurable, quantifiable piece of physical reality. The issue is not just sacredness. It is *sacred matter*. It is *sacred nature*.

The first task at hand, as we seek the road that biotechnology should take in the future, is to gain a foundational understanding of the concept of sacredness. This is the first thing that will be explored in this chapter. This task will begin by surveying the Biblical concept of sacredness, in an attempt to discover the core meaning of this religious concept as expressed in the both the Hebrew and Christian Scriptures. We will then observe some post-biblical developments in the understanding of the sacred, particularly in the Christian concept of *sacrament*. This will lead to the second major focus of this chapter, the intersection of the *sacred* and the physical. If a piece of nature, such as the ground upon which the burning bush existed or the bread of the Christian Eucharist, was deemed sacred, then does that open the road for all of nature to have the potential of being sacred? There are, therefore, two major questions to be explored in this chapter: What is the nature of sacredness; and what is the sacredness of nature?

A Biblical Concept of Sacredness

In seeking to establish the connection between the various areas of biotechnology and the prerogatives of God and humans within those areas, the place to begin is with the word that is so often used in debates and discourse. It is the word *sacred*. This word often becomes the trump card in these debates. Once the term is invoked, it seems that the lines have been drawn, and the complex issues of the ethics of biotechnology become as defined as black and white. Those who use this trump card, by defining a piece of biology (such as a gene or an embryonic stem cell) as *sacred*, postulate that this term means that these biological elements have an intrinsic divine value and therefore are completely restricted from human domination, manipulation, or destruction. Such a global definition of the term *sacred* leaves no room for negotiation. Those who are on the other side of the argument are left with no other strategy than to attack the very concept of sacredness, and strive to remove it from consideration all together. One side states that an embryonic stem cell is sacred; the other side argues that it is not. In such a discourse, there seems to be little room for the possibility that an embryonic stem cell can be both sacred and available for human manipulation or domination, albeit under certain conditions.

This common understanding of the term *sacred* makes it very difficult for people to hold a religious conviction based on the guidance of Scripture and at the same time, support the human domination of the biology of life. In a PBS interview with Dr. Ted Peters, the Professor of Systematic Theology at the Pacific Lutheran Theological Seminary in Berkeley, CA, Peters was asked to elaborate on his concept of the "Gene Myth". The following discourse occurred:

QUESTION: You've disagreed with this position that DNA is sacred.
DR. PETERS: Yes. I think what happened is that people began to treat DNA as sacred. By sacred I mean putting up no trespassing signs, saying you can't muck around with it, you can't get in with your wrenches and screwdrivers and mess around because DNA was put there by God. Well, I disagree with that. [9]

In another setting, a symposium at Harvard University in April of 2003, Peters was asked to consider the possibility, if hypothetical in his mind, that the gene was in fact sacred. As reported in *Science and Theology News*....

If it is sacred, he said, it should be left alone. In response to his own question, "Should we play God?" Peters said, "No, souls can't be Xeroxed. Souls are described to be dependent on divine responsibility alone, not the human genome." [10]

We will return to Peters' work later in this chapter. At this point, his words are offered as an example of the issue at hand. In many ways, the word *sacred* has become the sword of division between those who wish to prohibit many aspects of biotechnology and those who wish to proceed. In such a scenario of either/or, both sides are incomplete. Strict prohibitionists could benefit from a better understanding of the Biblical concept of sacredness, and those who wish to eliminate the idea of sacredness in order to open the door to biotechnology could benefit from seeing why such an approach is unnecessary, and is akin to tossing out the baby with the bathwater. Let us, then, proceed to discover the Biblical concept of sacredness.

Sacredness in the Hebrew Scriptures

As one begins an exegesis into this biblical word and concept, it is important to note that the word *sacred,* as it is so commonly used today, is a rather modern term. In many of our newer translations of the Bible, it is the oft chosen translation of a certain Hebrew word, along with its several derivatives. In older English versions of the Bible, the word *sacred* is not used. In the King James Version, for instance, the same Hebrew word is translated as *having been hallowed,* which moves closer to the concept of *holy.* In fact, the Hebrew concepts of *sacred* and *holy* are closely aligned. When Moses was told that he was approaching holy ground, it would be nearly an interchangeable phrase to say that he was approaching a "sacred space."

In Hebrew, the root word for *sacred* is קָדַשׁ (qodesh). As one surveys this word's appearance in the Hebrew Scriptures, the common interpretation of its meaning (that it indicates intrinsic divine value and therefore restricted from human intervention) immediately becomes suspect. This is due to the fact that the term is applied not only to God, or to Godly things, but many other things as well. In fact, there are some texts that associate the word *sacred* with ideas that are quite contrary to God. In Exodus 23:24, God is giving the Israelites instructions concerning the gods of surrounding tribes. The passage states

> Do not bow down before their gods or worship them or follow their practices. You must demolish them and break their *sacred* stones to pieces. [11]

It is also quite interesting to note that Hebrew terms for both female and male prostitutes, which are so often depicted in a negative sense, are derivatives of קָדַשׁ. [12]

What is observed, then, is that the Hebrew word that has been translated *sacred* can have many references. Many of these references are to God, or to Godly things. Religious practices, sanctuaries, priestly garments, and the like are all deemed as sacred. So too, however, are the stones of idols and the cultic prostitutes of pagan worship. If this is the case, then the common understanding of something *sacred* must be in error. Something sacred is not necessarily something that has intrinsic divine value. By *intrinsic*, we mean something internal and inherent, or as the Merriam Webster Dictionary states, *"belonging to the essential nature or constitution of a thing"* [13] If that is not what is intended by the word *sacred*, then what is? Numerous commentaries and Bible dictionaries all point to the same fundamental meaning that can be expressed in two central concepts. First, something sacred is something that has been *separated*. It is something that has been set apart, and therefore made distinct from other matter. The Interpreter's Dictionary of the Bible states that "the meaning of 'separation' is paramount ... the more elemental meaning seems to lie with 'separation'."[14] In his classic text entitled <u>The Sacred and the Profane</u>, religion historian Mircea Eliade describes the separateness of the sacred as a *"hierophany"*, which he describes as "an irruption of the sacred that results in detaching a territory from the surrounding cosmic milieu and making it qualitatively different."[15]

Along with the idea of separation, the second central concept is that this separated subject has been *dedicated* to a particular person or purpose. Often, this dedication represents an exclusive claim.

In the Hebrew Scriptures, then, we see an important distinction between our commonly accepted notion of sacredness and the actual Biblical meaning. Indeed, there are objects and spaces that are deemed sacred to God. The sacredness of them, however, is not something that they possess. It is not something that lies within them, and radiates out. Rather, sacredness is something that God imposes upon them, for whatever reason God may have. These spaces and objects are *made* sacred, and are thus separated out from all other matter and dedicated to God's special, and often exclusive, use.

This concept of the nature of the sacred is an important element of the theology of the ancient Israelites, and represents a pivotal departure from the religious beliefs of the other ancient peoples. The conventional religious thought of that time asserted that matter and space had intrinsic divine value, which existed from immortal time past and will continue for immortal time forward. Although the ancient Israelites did adopt into their theology a sense of physical matter serving as sacred space, there was one strong departure. It is what Brevard Childs, in his book <u>Myth and Reality in the Old Testament</u>, has entitled a "broken myth".[16] The ancient Israelites believed that physical matter could become sacred, but was not inherently sacred. In his essay entitled *Sacred Space, Holy Places, and Suchlike*, David Clines, of the University of Sheffield, states:

> The emphasis in the Old Testament is on sacred places as chosen by God rather than sacrosanct from of old. Of course, as we have seen, both these concepts are well attested in religions generally, but for the Old Testament, as far as I can see, the holiness of a place tends to be a quality acquired through becoming in history a place of divine manifestation rather than an inherent quality it has had from primaeval times. Zion is not a holy place since the Urzeit, but has become a holy place-for Israel-with experienced time. Of course its holiness is older than David's time, but the Hebrew story is that the site of the temple was no sacred spot hallowed from time immemorial, but a place originally profane: a threshing-floor belonging to a Jebusite (2 Sam. 24.15-25; 1 Chron. 21.15-16). It is not Zion, but God's heavenly throne, that is established from of old (Ps. 93.2). [17]

This clarification of the meaning of the word *sacred* has significant implications for this study, and for the whole debate concerning the ethics of biotechnology. Before the implications of this subject are articulated, let us first examine the usage of the term *sacred* in the Greek New Testament, to see if this understanding of the term is in any way altered or modified by it.

Sacredness in the New Testament

Before getting into the New Testament proper, it must first be recognized that the theological terminology that is found there is very much connected to the Greek version of the Hebrew Scriptures. This Greek version, known as the Septuagint, was written hundreds of years before the New Testament documents were created. Therefore, when a New Testament author used the Greek word for *sacred*, that author carried forward the meaning of that Greek word from its Septuagint usage, and would intend a directly correlated meaning with the Hebrew term קָדַשׁ.

The Greek translation of that Hebrew word is ʿαγιαζ, and is translated by some English versions as *sacred* and by others, *holy*. This again illustrates the close proximity of the meaning of these two words. In the New Testament, the word has a much more limited use than in the Hebrew Scriptures. One distinction that can be made is that there seems to be no New Testament occasion in which something associated with idolatry or sinful practices is qualified with the word sacred. Even so, it would be wrong to suggest that the New Testament applies any sense of intrinsically divine value upon objects deemed sacred. One particular passage in the book of Romans makes this clear:

> One man considers one day more sacred than another; another man considers every day alike. Each one should be fully convinced in his own mind.[18]

In this case, it is clear that the term *sacred* does not imply an intrinsic value. Whether or not the day is sacred is up to the observer. It seems clear that the application of the word 'αγιαζ in this passage continues the meaning that was observed in the Hebrew Scriptures, that of indicating something that is separated, set apart, or perhaps, made special.

Although the New Testament offers an interpretation of the word *sacred* that continues the concept observed from the Hebrew Scriptures, it does broaden significantly the application of that concept in terms of what can, in fact, be sacred. Christian theology introduced the concepts of the Incarnation, of the Church as the body of Christ, and of the sacraments.

Sacramental theology, which finds its primitive roots in the New Testament but has received exhaustive treatment in all the years since, will become a very important ingredient in this book, and will be more fully explored later in this chapter. For now, however, it is highlighted as a poignant example of the relationship between physical matter and the sacred. There is, of course, no unified understanding of what transpires at the table of Holy Communion, when the bread and the wine become to us the body and blood of Christ. The Roman Catholic tradition holds to the theology of *transubstantiation*, in which the bread and wine become the body and the blood. Martin Luther introduced the belief in a *consubstantiation*, in which the elements co-exist as both bread and wine as well as the body and the blood of Christ. John Calvin, from whom my own theological tradition has evolved, rejected both ideas. He believed that the bread and wine remain as such, but in the sacrament Christ becomes spiritually present. It is not the purpose here to argue one view over the other. All three views, however, carry one central concept concerning the physical elements brought to the table. In all three views, the bread and wine, as elements of the physical world, have no pre-existing, or intrinsic, sacredness. They only become sacred when God sets them apart (separation) and chooses them for God's special activity (the presence of Christ).

We will return to sacramental theology in the second main focus of this chapter, in which the question of whether nature is sacred to God will be addressed. At this point, however, it is time to synthesize this chapter thus far with the main questions of this book. Now that the meaning of the term *sacred* in the Hebrew and Christian Scriptures has been explored, it is time to articulate the implications for the ongoing debates over the appropriateness of the controversial aspects of biotechnology.

The Implications

At the heart of the debates and controversies concerning biotechnology, there seems to be one core question. *Is it sacred*? Is DNA sacred? Are embryonic stem cells sacred? Is biological life sacred? These are obviously religious questions, even though they seem to exercise much control and influence in the public and political debates surrounding scientific enterprise. When these important questions are considered from a particular religious tradition, it is important that one maintains congruency with the guiding theology of those traditions. In the current political and religious discourse over biotechnology and the sacred, it seems that the predominant understanding of the *sacred* is more akin to the primitive religions over against which our Judeo-Christian tradition emerged than from the tradition itself. Religious and political leaders speak of physical matter as being sacred, as if that means an inherent quality that always was and always will be. The Biblical tradition, however, resists that notion. Physical matter is not sacred in and of itself. It is only sacred if and when God chooses to separate it out, and claim it as God's own territory of use and purpose.

Given this understanding, this book begins with a claim that may sound radical, especially coming from a theological mindset. Is it sacred? The answer is *no*. DNA is not sacred. Embryonic stem cells are not sacred. The separation of human life and animal life is not sacred. A biological arm is not any more sacred than a mechanical one. Neural cells are no more sacred than digital components. Carbon based intelligence is no more sacred than silicon based intelligence. The biological existence of carbon based life is no more sacred than an existing mind uploaded into cyberspace.

The list could go on, but we will stop here to make the point. Whenever one asks the question "Is it sacred?", and stops there, the answer must always be *no*. Asked in this way, the object of the question assumes an internal and intrinsic quality. Such an internal and intrinsic quality is not in keeping with the Biblical understanding of sacredness. If one wishes to keep aligned with that Biblical understanding, then one must recognize that physical matter can only be sacred if it is in relationship with the Other, which is God.

We must clearly delineate the physical matter from the One who makes things sacred, or we risk merging the two with a resulting theology that resembles pantheism more than a Judeo-Christian faith. As Martin Buber's classis book I and Thou reminds us, keeping God a "Thou" to the world's "it" is of critical importance. Buber writes

> By its very nature the eternal You cannot become an it;
> And yet we reduce the eternal You ever again to an It, to something, turning God into a thing, in accordance with our nature.[19]

Whenever one looks at a piece of physical matter as if it has an inherently divine quality, whether it is the mythical Holy Grail or a piece of basic biology under a microscope, he/she is wandering off the path of a Biblical understanding of sacredness.

It has been articulated, then, that the oft asked question *"is it sacred?"* must always be answered in the negative. This does not mean, however, that biotechnology should be free from all theological and moral constraints. This is, obviously, only the beginning of this journey. In fact, the above question is not so much wrong, as it is incomplete. If physical matter does not have an intrinsic sacredness, then it only becomes sacred if God chooses to separate it out and declare it for God's will and purpose. "Is it sacred?" is a misguided question. "Is it sacred *to God?*" is a whole other question, and to borrow a cliché, it is a whole new ballgame. Asked in this way, one may find entirely different answers.

Some may be inclined to see the difference between these two questions as splitting hairs. Actually, however, there is a significant difference. To say that something has intrinsic divine value is a very different claim than to say that its divine value is given to it by an outside entity, namely God. The differences between the two will be explored in the conclusion of this chapter. First, however, there is a very large question to tackle. *Is nature sacred to God?* It is to this important question that we now turn.

Nature and the Sacred

In order to address this important question, one must begin by establishing a working definition of what is meant by the word *nature*. It may seem simple, but with a closer look it becomes apparent that the common understanding of *nature* has experienced a significant shift over the last century or so. This shift has largely contributed to the complexity of the ongoing debates and controversies concerning biotechnology.

Toward a Definition of Nature

In a basic sense, nature is the physical world in which we "live and move and have our being."[20] Historically, we have known nature by a theologically laden term, *creation*. Within that word, however, there is the sense of totality that exists within the concept of nature. Nature encompasses almost everything. In <u>On the Moral Nature of the Universe</u>, co-authors Nancey Murphy and George F.R. Ellis build a moral view of cosmology beginning with this working definition of nature:

> The foundation of our understanding of cosmology is our view of nature, understood in terms of matter, whose behavior is determined (or at least accurately described) by casual laws.[21]

Arthur Peacocke, in his book <u>Theology for a Scientific Age</u>, offers a similar yet expanded definition that is useful for us. He writes:

> In the Newtonian perspective, which dominated the mind of the West for two and a half centuries-and still prevailsin the general intellectual climate since it corresponds so well with

18

intuitions derived from our senses-the stuff of the world, *matter,* possesses *energy,* and is located in *space* at a particular *time.*[22]

Matter, energy, space, time…these are the elements of nature. It is quite observable that these are also the objects of science. These are the items that can, albeit with certain limits, be measured, quantified, defined and within our limited abilities, dominated.

A Shifting View of Nature

For many centuries, humans have been dominating nature. To a large degree, this has been without any large movement of religious opposition until recent history. One can identify some basic reasons for this, which will lead to understand better why the current sciences of biotechnology seem to collide so strongly with religious convictions. In the following section, we will examine two developments in the evolving understanding of and relationship to nature. These are 1) a shifting worldview; and 2) humanity's place in nature.

A Shifting World View

First, over the last several centuries of scientific advances, there has been a dominant *world view,* in which nature (as it was understood) was set free from ancient mythology. That ancient mythology, as already noted, believed in the intrinsic sacredness of certain physical matter. Leonardo Boff, a scholar of the Franciscan Order, called this the *theocentric cosmology,* in which nature was seen as a "hierarchical, sacred, and unchangeable whole"[23] With the arrival of the Enlightenment, this ancient world view was replaced by the one that has dominated the world, and particularly the West, until our current age. Boff described this era as the *anthropocentric cosmology,* in which there was a belief in a dual cosmology of matter and spirit. According to Boff, this has been an age of "arrogant anthropocentrism", in which the world is reduced to natural resources for human consumption.[24] This theme is certainly echoed in the classic essay by Martin Heidegger entitled *The Question Concerning Technology,* in which we read…

It remains true, nonetheless, that man (sic) in the technological age is, in a particularly striking way, challenged forth into revealing. That revealing concerns nature, above all, as the chief storehouse of the standing energy reserve.[25]

In this world view, there is a clear separation between matter and spirit. Nature is clearly in the realm of matter, and is therefore the *standing reserve* for human domination. This world view, whether it is called "the scientific age", or (per Heidegger) "the technological age", or (per Boff) "the age of the anthropocentric cosmology", has been so deeply entrenched that it may be taken for granted as a lasting reality. A more careful examination of our time, however, may reveal that this world view is cracking, and a new one is emerging.

Since we have invoked Arthur Peacocke's definition of nature, it should be made clear that he offered those words as a precursor to suggest that this view of nature is experiencing a fundamental shift. He goes on to say

It may be roughly characterized by the assertion that the understandings of space, time, matter and energy of the 'classical' perspective turn out to be approximations of more subtle, and often unpicturable, concepts and entities which are related in ways that were unthinkable before the first decades of this century (20 C). For during that period relativity theory, in its special and general forms, and, even more iconoclastically, quantum theory, together caused a complete revolution in human understanding of the physical world, the consequences of which are still being absorbed into philosophy- and hardly yet into theology.[26]

Indeed, the long held sense of nature is undergoing a major transition. With the emergence of such theories as parallel universes, string theory, and multiple dimensions, the traditional understandings of space, time, energy, and matter are coming up short. Today, the hard sciences are being overshadowed by metaphysics, which begins to challenge the long held notion that science can explain all things. This new sense of the limits of science has opened the door to the emergence of a new world view. Many have named this new world view as *post-modernism*. In his book <u>Rebuilding the Matrix</u>, Denis Alexander writes:

if modernism is characterized by the 'standard' view of science…, enthroning the scientific method as an arbiter of what is rational, then postmodernism proclaims that science provides only one (among many) culture-bound ways of looking at the world. Science may thus be treated as one option on the worldview shelf displayed by multicultural societies in which occult or mystical worldviews may be looked on as equally valid.[27]

The ambiguity of a hard and solid definition of post-modernism is illustrated by the fact that it can only be defined in relationship to what it is supplanting. We may not know yet what new world view will take firm hold in the 21st century, but it seems certain that it will be different than the familiar scientific age. Whatever this new world view may become, it seems quite certain that it will offer a shift in the understanding of nature. Leonardo Boff, to whom we have looked to get a sense of our past world views, offers his view of this emerging age. Matthia Kadavil describes Boff's thoughts this way:

Boff delineates a new cosmology that developed as the outcome of the relativity theory of Einstein, the quantum physics of Bohr, the intedeterminacy principle of Heisenberg, the theoretical physics of Prigogine and Stengers, the depth psychology of Freud and Jung, the transpersonal psychology of Maslow and P. Weil, as well as biogenetics, cybernetics, and deep ecology. Boff also calls attention to the emergence of a new religious dimension grounded on this new cosmology. It is the experience of the divine as a globalizing phenomenon. An important characteristic of this experience of globality is that it can discern the presence of God both in the secular and the sacred. Moreover it contains an integrative tendency present in all dimensions.[28]

Kadavil's analysis leads into an important discussion of the bridging of the sacred and the secular, and we will return to his thoughts momentarily. The point thus far, however, is to seek an understanding of this idea called *nature*. Initially, it may seem easy to define nature. Yet, it has been seen that the initial definitions may be presumptuous in relationship to a particular world view that is

21

undergoing changes. Traditionally, we have seen nature as physical matter that is sharply delineated from the spiritual. Whatever postmodernism is, it seems evident that this delineation is not such a given. The existing tensions revolving around aspects of biotechnology may not only be due to the new territories that biotechnology is conquering, but also to a shifting worldview that interprets nature in a whole new way. For some, this is a welcomed advancement, particularly among those who hold a passion for ecology. In an article entitled *The psychopathology of the Human-Nature relationship*, Ralph Metzner writes of this new worldview...

> The natural is regarded as the realm of spirit and the sacred. The natural is the spiritual. From this follows an attitude of respect, a desire to maintain a balanced relationship, and an instinctive understanding of the need for considering future generations and the future health of the ecosystem-in short, sustainability. Recognizing and respecting worldviews and spiritual practices different from our own is perhaps the best antidote to the West's fixation on the life-destroying dissociation between spirit and nature.[29]

Humanity's Place in Nature

This shifting definition of nature is not only derived from the fields of metaphysics, but also from Darwinism and molecular biology. Just as the traditional world view has offered a clear distinction between spirit and matter, so too has it offered a clear distinction between humanity and nature. The Judeo-Christian faith tradition has certainly understood that humanity was, in one way, a part of the realm of nature. The creation story recorded in Genesis 2 teaches that humanity was formed from "the dust of the ground", and the image is somberly invoked during the pronouncement of judgment after the fall, in which God tells Adam "From dust you are, and to dust you will return" (Genesis 3:19, NIV).

Even though, however, the traditional beliefs have affirmed humanity as a part of nature, they have also enforced the notion that humanity is special, and granted a hierarchal position above nature. This has created the sense that humanity is somehow distinct from nature, and therefore relates to it from an outside-nature posture. In the Genesis account of creation cited above, the human (Adam) is created, and then placed *into* the garden, which represents nature.

This image reinforces the idea that human life is distinct from nature. Furthermore, humanity is created "in the image of God", which seemingly endows humanity with a special quality over and above nature. John Calvin expressed it this way:

> For although God's glory shines forth in the outer man (sic) yet there is no doubt that the proper seat of his image is in the soul. I do not deny, indeed, that our outward form is so far as it distinguishes and separated us from brute animals, at the same time more closely joins us to God. And if anyone wishes to include under "image of God" the fact that, "while all other living things being bent over look earthward, man has been given a face uplifted, bidden to gaze heavenward and to raise his countenance to the stars," I shall not contend too strongly-provided it be regarded as a settled principle that the image of God, which is seen or glows in these outward marks, is spiritual.[30]

Once humanity is seen as having a qualitative difference from the rest of nature, it follows that humanity would begin to relate to nature as something different. If we may utilize the language of Martin Buber once again, to isolate one species (humanity) and assign it individuality it then becomes an "I", which necessitates an "It". Buber expresses this by saying:

> *There is no I as such but only the I of the basic word I-You, and the I of the basic word I-It.* [31]

Once this separation between humanity and the rest of nature was established in Judeo-Christian thought, it led to two streams of thought and action. The first is the notion that nature is there for human consumption and domination, the *standing reserve* of Heidegger's essay. Given this stream of religious consciousness, plant and animal life is totally exploitable for human benefit. Perhaps this is why we are more willing to accept the cloning of animals, for example, than humans. We will return to this theme in Chapter Four.

The second stream that has developed from our theological understanding of humanity is quite the opposite. It is expressed in the word *stewardship*. In this stream of religious consciousness, humanity is charged with the management and conservation of nature's resources. In The Christian Vision of Humanity, John Sachs states:

> We are called to a cosmological and ecological mutuality which is founded on the goodness of creation and the delight which the Creator has in it. Therefore, to be God's image or representative on earth, to share in God's dominion, means that we receive a share in God's power *for* creation, not simply *over* creation. It does not give us a license to exploit it as we please. Human beings are accountable to the Creator for the world's well-being and wholeness.[32]

Over the course of the centuries, our faith tradition has supported a certain view of nature, and of humanity's particular relationship to it. Whether it has led to exploitation or conservation, the central concept has been at work. Humanity is different. Humanity has been created in God's image. Humanity shares a biology with nature, but is still somehow an "I" to nature's "It."

This view of a human separateness from nature has been seriously eroded over the last several decades. This erosion began with the groundbreaking work of Charles Darwin, whose 1859 release of Origin of the Species, and even more so, his 1871 release of Descent of Man, offered a new proposal of humanity's place in nature. Humanity was not distinct and special. Humanity evolved just like every other species. In fact, humanity has shared a common ancestry with existing primates. In our own time, Darwin's incomplete theory of descent has been greatly enhanced by the synthesized theories of Neo-Darwinism.[33]

It would be easier to meld this information into our traditional understanding of anthropology if the implications of this science applied only to our "physical" bodies, over against our understanding of "soul" or "spirit". Yet more and more, molecular biology is offering up evidence that even the areas we consider to be non-biological are, in fact, related to the inner workings of our DNA. In recent years, pioneering work has been done on the biology of intelligence, personality, emotional attachment, and even spirituality itself. Dean Hamer shocked the world when he introduced the "gay gene", although his research has since been discredited.[34] That did not stop him, however, from making another dramatic claim. He had identified the VMAT2 gene, which he labeled "*The God Gene*":

The VMAT2 gene variant containing a C-or 'spiritual allele', as I began to think of it-was present on only 28 percent of chromosomes, compared with 72 percent carrying an A. But because both the C/C and C/A genotypes had increased self-transcendence scores, compared to the A/A genotype, it worked out that 47 percent of people in our study were in the higher spirituality group, as compared to 53 percent in the lower group-virtually half, which was exactly what we were looking for. While this one gene may not make one a saint, a prophet, or a seer, it was enough to tip the spiritual scales and predispose one toward spirituality.[35]

With molecular biology research touching the most "spiritual" aspect of human life, the very core of our traditional understanding of humanity in relationship to nature has been challenged. Now, nature is no longer the "It" to our "I". We are, in every sense of being, part and parcel to nature. For some, this has led to a bold new approach to human life, in which the traditional sense of human separateness from the rest of nature is no longer seen as a protected value. Perhaps the most poignant example of this is offered by the ongoing medical research that utilizes transgenic animals that have been given human molecular materials, thus crossing the often considered "sacred" line between humanity and other biological life. In the November 20, 2004 edition of the Washington Post, an article entitled *Of Mice, Men, and In Between* reported the research activities of Irving Weissman, director of Stanford University's Institute for Cancer/Stem Cell Biology and Medicine. The story reported that Weissman's team, who had already created a mouse with a nearly complete human immune system, was now injecting human neural stem cells into mouse fetuses. The hope is that such human brain tissue in mice would offer a vehicle for finding new cures for neural diseases. The article concludes with an interesting possibility:

> Weissman says he is thinking about making chimeric mice whose brains are 100 percent human. He proposes keeping tabs on the mice as they develop. If the brains looks as if they are taking on a distinctively human architecture-a development that could hint at a glimmer of humanness-they could be killed, he said. If they look as if they arc organizing

themselves in a mouse brain architecture, they could be used for research.[36]

This is but one example, poignant as it is, of the blurring lines between humanity and the rest of nature. When EB White wrote his novel <u>Stuart Little</u> back in 1945, his humanlike mouse was clearly a fictional character. In the brave new world of biotechnology, yesterday's fiction may become tomorrow's fact.

While the new age of molecular biology, which has brought humanity fully into the realm of nature, has allowed some to feel a new freedom of domination over human life, it has caused others to feel a heightened sense of anxiety and fear. A generation ago, CS Lewis expressed this fear in his book <u>The Abolition of Man</u>:

> The wrestling of powers *from* nature is also the surrender of things *to* nature. As long as this process stops short of the final stage we may well hold that the gain outweighs the loss. But as soon as we take the final step of reducing our own species to the level of mere Nature, the whole process is stultified, for this time the being who stood to gain and the being who has been sacrificed are one and the same.[37]

Whether or not our journey of biotechnology will lead to the "abolition" of humanity is yet to be seen. Given the convergence of that technology with natural selection's revision of humanity's place in nature, CS Lewis could very well be right. If humanity is nothing special, there is no guarantee that there will be a permanent place for us in this world. If we succeed at creating intelligent machines, we may find that they will overtake us in the race of the survival of the fittest. If we succeed at replacing biological and neural components with mechanical and digital ones, we may find that human consciousness can exist without a biological human body. We have not yet seen CS Lewis' "abolition of man (sic)", but it certainly can be argued that we are witnessing the abolition of our Judeo-Christian anthropology. Humanity is not *apart* from nature. Humanity is *a part* of nature. Even as the Judeo-Christian anthropology wrestles with this realization, some still cling to the belief that humans have, among all other entities of nature, a particular relationship with God. Noreen L. Herzfeld, in her book entitled <u>In Our Image: Artificial Intelligence and the Human Spirit</u>, states that in spite of the possibilities of creating an intelligence that is artificial, "the image of God in humankind has become one cornerstone of Christian anthropology, a locus for understanding who we are in relationship to both God and the world."[38]

In this chapter, we have been wrestling with two important concepts that are foundational to the rest of this book: Nature and the Sacred. We first developed a contextual understanding of the word *sacred* as it is inherited from the Hebrew and Christian scriptures. It was concluded from that study that sacredness is not an internal, inherent quality of physical matter, but rather a quality given to it by God. It is the quality of being set apart for God's unique purpose and will. Given this understanding, we realized that the proper question of any physical matter is not "is it sacred?" but rather "is it sacred *to God*".

Attention was then turned to the second main concept, that of nature. We began with a simple definition that stated that nature represents matter, space, time, and energy. From that starting point, however, we recognized two important evolutions that are occurring in the current understanding of nature. It was first recognized that the new theories of physics and metaphysics have caused us to begin to think of reality that is beyond the measurements of nature, and has led to the emergence of a new world view that offers possible new interpretations of nature. Secondly, it was recognized that the long held belief in the "otherness" of humanity in relation to nature has been largely unseated. As humans, we are an integral part of this thing called *nature*. The world view of postmodernism, the new questions emerging concerning religious anthropology, and the unprecedented advancements of biotechnology coalesce into the perfect storm of controversy that so defines our times.

Now that an understanding of the words *sacred* and *nature* have been articulated, it is time to bring them together to address an important question for this study: *Is nature sacred to God?*

Is Nature Sacred to God?

Asked in different eras of time and culture, this question would receive different answers. In primitive times, the answer was clearly and unequivocally answered in the affirmative. In describing the way that ancient cultures would answer this question, Mircea Eliade states:

Nature is never only "natural"; it is always fraught with a religious value. This is easy to understand, for the cosmos is a divine creation; coming from the hands of the gods, the world is impregnated with sacredness. It is not simply a sacrality communicated by the gods, as is the case, for example, with a place or an object consecrated by the divine presence. The gods did more: they manifested the different modalities of the sacred in the very structure of the world and of cosmic phenomena.[39]

Eliade goes on in his book to describe the "desacralization of nature" that has been such a hallmark of the scientific age. As has already been described, this has been an age in which a stark wall of division has been erected between the realm of the spirit and the realm of matter. During this era of human development, nature has been stripped of its sacred potential. In addition, the age of astronomy and physics has taken what was previously seen as totally spiritual space and brought it into the realm of mundane matter. One must consider what a contest of faith that would be to people of religious faith in order to more fully appreciate the church's position in the Galileo affair.

It seems almost ironic that in the very age in which we are poised to cross new boundaries in our mastering of the elements of chemistry and biology, there is also emerging a new world view in which the long standing wall between nature and the sacred is disappearing. As was witnessed in the words of Leonardo Boff, this new world view is a fertile field for a reunification of nature and the sacred. As will be shortly seen, Boff is one theologian who has decided to plant an idea in that fertile field, and that idea is growing. It is a theology that is mainly coming from the East, and from Orthodox Christianity. It is a theology known as *The World as Sacrament.* Before we consider this idea, let us first chart a course that will take us to its doorstep.

Is nature sacred to God? What does that question really mean, in light of our research? It means, is there any evidence that God has separated nature out, and claimed it for God's own purpose? It does seem that such evidence exists, and does so in one of the most basic doctrines of the Christian faith. It is the doctrine of *general revelation*. The theology of general revelation asserts that God reveals God's self and God's divine purposes through the elements of nature. This theology finds its origins in both the Hebrew and Christian

Scriptures. The Psalmist proclaims: *The heavens are telling the glory of God; and the firmament proclaims his handiwork.*[40] The letter to the Romans in the New Testament says: *Ever since the creation of the world his eternal power and divine nature, invisible though they are, have been understood and seen through the things he has made.*[41]. John Calvin, in his Institutes of Christian Religion, explains that nature (i.e. creation) exists for a definite purpose, for "it was his will that the history of Creation be made manifest, in order that the faith of the church, resting upon this, might seek no other God but him who was put forth by Moses as the Maker and the Founder of the Universe"[42].

The doctrine of General Revelation asserts that all of nature has been chosen (separated out) by God's own purpose, which is to show humanity God's own person and God's own nature. Of course, to suggest that all of nature has been separated out leaves a fundamental question: from what has it been separated? To say that everything is special makes nothing special. To say that all of nature is sacred seems to suggest that none of it is. If nature is sacred, and to be sacred means to be distinguished and separated from all else, what is the "all else"? Perhaps the only possible answer is: We don't know. It is intriguing to think, however, that the emerging theories of parallel universes and multiple dimensions may be giving us our first glimpse into the possible space beyond the nature that our faith teaches has been made sacred by God. In the strange theories of Metaphysics, are we beginning to discover nature's "otherness"?[43]

General Revelation, then, teaches that God has used nature to reveal God's self to humankind. The discovery of God is a discovery of grace. Therefore, nature does more than just point us to God. Nature itself becomes a vehicle by which one encounters God, and through which one can find grace. If a piece of physical matter (i.e. nature) is not only being set apart by God for a special purpose, but actually becoming a means of transmitting grace, we have stumbled onto something. It is the Christian definition of the word *sacrament*. As the Church proceeded to institutionalize the theology of sacrament, it focused so strongly on the most commonly accepted modes of sacramental grace, chiefly the Eucharist, that it lost the sense of nature's ability to serve as sacrament. In the 20th century, Christian thinkers began to revisit this idea that nature itself possesses sacramental qualities. This was not so much a new theological premise, as it was a rediscovery of concepts that were present in the earliest days of the Church. Kadavil writes:

> The Bible and Revelation are the first source of God's revelation prior to the historical incarnation of Jesus. It leads us to creation, which is the beginning of God's presence in the world. At the very outset it should be made clear that going back to the Creation is not something new, rather it is the result of our ongoing effort to rediscover the earlier understanding of sacraments.[44]

The notion of the world as sacrament has largely emerged from the Orthodox faith. It was introduced as a developing theology by Orthodox theologian Alexander Schmemann in 1964.[45] Since then, numerous Orthodox theologians have contributed to this theme. More slowly, some Catholic theologians have advocated this theology. Among them is the one with whom we began this discourse, the Franciscan theologian Leonardo Boff. Boff has been a significant contributor to this idea, although he is mainly known for his leadership in Liberation Theology. From within the Protestant tradition, we find some movement toward this theme in Jurgen Moltmann's book <u>The Church in the Power of the Spirit</u>, in which Moltmann illustrates that the activity of God's Spirit should be seen as sacrament.[46] Even closer to the theme of *world as sacrament*, consider Langdon Gilkey's <u>Nature, Reality, and the Sacred</u>, in which he concluded his entire book with these words:

> To know God truly is to know God's presence also in the power, the life, the order, and the redemptive unity of nature. Correspondingly, to know nature is to know its mystery, its depth, and its ultimate value-it is to know nature as an image of the sacred, a visible sign of an invisible grace.[47]

"A visible sign of an invisible grace" is the definition of a sacrament.

Although one can thank the Orthodox faith for getting this dialogue started, one may wonder why it was that the Orthodox tradition more easily accommodated this important idea. Kadavil suggests that *the world as sacrament* idea can more easily fit into the Orthodox theology because it has a more open sense of sacrament, based on the Greek word μυστηριον (mysterion). This belief in sacraments has no limit on the number of possible sacraments; so that the door is wide open for all of nature to be considered. The Catholic idea of sacraments, based on the teachings of Peter Lombard[48], limited the possible number of sacraments to the seven of which they were well aware. This left little room for a widened belief in sacraments, especially wide enough to consider all of nature.

There is one other difference between the orthodox and catholic beliefs concerning sacraments that must be addressed here, mainly because it seems at first to contradict a major focus of this chapter. In the Catholic faith, the sacraments become sacred when they are properly administered by the actions of a priest. In Eastern Orthodoxy, the *mysterion* can exist in things that have had no such priestly oversight. In explaining this ability of nature to be sacramental without liturgical practice, *the world as sacrament* theologians talk of nature as having an "intrinsic" sacredness. Philip Sherrard[49], in an essay included in The Orthodox Ethos: Essays in Honour of the Centenary of the Greek Orthodox Archdiocese of North and South America, writes:

> Sacrament is not something set over against, or existing outside, the rest of life, so that it is sacred while the rest of life and all other things are non sacred or profane or non-sacramental; it is not something extrinsic and fixed in its extrinsicality, as if by some sort of magical operation or *Deus ex machina* the sacramental object is suddenly turned into something other than itself and different from all other created objects. On the contrary, what is indicated or revealed in the sacrament is something universal, the intrinsic sanctity and spirituality of all things, what one might call their real nature.[50]

This language of an "intrinsic sanctity" seems to contradict our earlier descriptions of the concept of sacredness in the Judeo-Christian experience. Earlier in this chapter, it was asserted that within this tradition, physical matter does not have an intrinsic sacredness; it is made sacred by God as a chosen vessel of God's grace. Although there seems to be here a contradiction, it is important to note that there is a difference in meaning between the two. Sherrard is making the claim that nature can be sacred without any human priest invoking sacredness upon it. Even so, he would agree that nature receives it sacredness from God. As such, the question remains the same as it has been posed in this study: not "is nature sacred", but rather, "is nature sacred *to God*?" One can observe this in Sherrard by looking at the following quote:

> Everything is capable of serving as the object of the sacrament, for everything is intrinsically consecrated and divine; is in fact, intrinsically, a "mysterium".[51]

Sherrard's use of the word "consecrated" illustrates that he sees an act of God upon the object in order to deem it sacred. That act, however, is not through the words of the priest, but at the very moment of the object's creation. This is what Sherrard, speaking for the Orthodox tradition, means by the notion of intrinsically. Something made sacred at its moment of creation (even made sacred *by* its creation) does not mean that it is sacred in and of itself. To further elaborate this important distinction, the reader is encouraged to consider the following two quotes of Vladimir Lossky, who wrote in his book The Mystical Theology of the Eastern Church:

> The creature itself is thus, by virtue of its very origin, something which changes, and is liable to pass from one state to another. It has no ontological foundation either in itself (for it is created from nothing) nor in the divine essence, for in the act of creation God was under no necessity of any kind whatever.[52]
>
> The tradition of the Eastern Church knows the creature tending toward deification, transcending itself continually in grace. It avoids, however, the attribution of a static perfection to created nature considered in itself. For that would ascribe a limited fullness, a natural sufficiency, to beings which were created that they might find their fullness in union with God.[53]

In response to the question, *is nature sacred to God*, we have now come to the answer. To be sacred, something has to show evidence of being set apart and dedicated for God's own purpose. Through the doctrine of General Revelation, one can observe that indeed, nature does show evidence of being set apart in such a fashion and for such a purpose. Nature is, in fact, sacred to God. Beyond that, however, we have discovered a stream of Christian thought that suggests that nature is not only sacred, but is actually a sacrament. This means that nature does not only exist to reveal God to humankind, but is actually meant to be a source of God's grace to humankind. This notion that nature is meant to be a source of grace will have an important voice in questions concerning the use of biotechnology. It should kept in mind, for we will certainly return to it again as we proceed.

In this first chapter, then, we have discussed two important concepts that are foundational to this study. We have considered the nature of sacredness, and the sacredness of nature. The concluding section of this chapter will synthesize our findings into a proposal of how one can proceed to discover a proper response to the questions of biotechnology.

Chapter Conclusions

This chapter, and indeed this entire book, began with a very important question for our time. Is it right for humans to proceed with the domination and manipulation of nature through the growing abilities in biotechnology? At the heart of this question, one finds the fundamental issue of sacredness. We have witnessed three different views of the sacredness represented in biology, and each one leads us to a different path in our exploration of the biotech frontiers. A helpful analogy to assist in this conclusion, and one that fits well when considering a road to travel, would be the common traffic light. The traffic light offers three possible instructions, represented by lights that are red, yellow, or green. Each of the three possible views of the sacredness of biology corresponds with one of those lights. First, we will identify the views, and the corresponding lights, that have been dismissed in this study. That will leave us with the one view, and the one light, that will lead us on down the road.

The first view, commonly represented by those who object to biotechnology, asserts that certain biological aspects have an inherent and intrinsic sacredness, as if they are sacred in and unto themselves. If this is the case, then the light is red. A red light means *stop*...you cannot proceed. If a basic element of biology, DNA for example, is intrinsically and inherently sacred, then its sacredness is fixed. Sacredness becomes its permanent feature.

The second view, represented to some degree by Ted Peters, is to abandon the notion that biological elements are in any sense sacred.

Peters particularly addresses the alleged sacredness of DNA in his arguments against the "gene myth". Peters writes:

> There is something special about DNA, to be sure; the particular genome that DNA bequeaths to each of us is largely determinative of our individual identity. Yet this is insufficient reason for treating it as functionally sacred.[54] (Peters, p.14)

In a similar vein, Robert Cole-Turner has argued for the de-sacredness of DNA this way:

> Such a conviction of DNA's sanctity is not grounded in western philosophical or religious traditions. Employing it now, in the context of genetic engineering, is arbitrary. To think of genetic material as the exclusive realm of divine grace and creativity is to reduce God to the level of restricting enzymes, viruses, and sexual reproduction. Treating DNA as matter-complicated, awe-inspiring, and elaborately coded, but matter nonetheless-is not in itself sacrilegious.[55]

In this view, there is a clear separation between the concepts of matter and of sacredness. As such, it is aligned with the scientific world view that has dominated our culture for a long time, but is now under growing question. If matter is separated from sacredness, then the light is green. Humanity has the *right* to proceed. I have emphasized the word *right* for a purpose. A right represents an inalienable freedom. If humans have a right over nature and matter, then we can proceed with no limits or parameters, save those that we establish by our own human consensus. Ted Peters, Robert Cole-Turner, and the like are certainly not advocating an ethic-free approach to biotechnology. Peters advocates for a science "in the service of beneficence" which leads to the exercise of biotechnology "in a free and responsible way."[56] Yet, to divorce nature from sacredness gives humanity an absolute right to master nature. Biotechnology seen as an absolute human right can easily lead to actions which Peters himself would deem irresponsible.[57]

We have witnessed the red light and the green light, both of which have been brought into question in this book. We have seen that nature does not have an intrinsic and inherent divine quality.

Sacredness only happens in relationship to God, as God sets something apart and claims it for God's purpose and activity. The red light view of biotechnology is not the chosen approach. We have also seen, however, that there is good reason to believe that God has set nature apart, and claimed it for God's purpose and activity. Nature is God's vehicle of general revelation, and serves as God's sacrament. Nature is sacred to God. As such, humanity does not have an inalienable right to master it. The light is not green.

If the light is not red or green, that leaves us with one other choice. It is the yellow light. The yellow light instructs us to proceed with caution, and to prepare to stop if necessary. If God is the one who determines the sacredness of nature, then the *right* to master nature belongs exclusively to God. That does not mean, however, that there is a permanent and lasting "no trespassing sign" posted over the natural realm. If God has the exclusive right to sacred space, then God has the right to extend *permission* for others to enter it. *Permission* is a much different realm of operation than *right*. In a right, freedom is inalienable. In a permission, freedom can be stipulated. As in a yellow light, we can proceed; but we must do so *with caution*. It is a stipulated freedom.

It is not a new concept for biotechnology to be governed by a guiding precautionary ethic. In the pre-war years of Germany, a legal philosophy known as *vorsorgeprinzip* (in English, "precautionary principle") was developed.[58] Since then, this principle has become an oft-applied concept to issues of environmental impact. The precautions of this principle are almost entirely based on matters of the safety of a proposed action vs. the risk of the contamination of the environment. In this book, the locus of precautions does not originate in questions of environmental safety, although that will certainly play a part. Rather, the locus is placed in the character of the one who has yielded sovereign authority to humans, namely God.

This chapter began with the image of Moses, as he had his famous encounter with the burning bush. In that passage, we discovered the existence of *holy ground*. It was sacred matter. God, and God alone, had the exclusive right to operate there. Yet, God exercised God's right to extend permission to Moses to enter that space. With that permission, however, came certain stipulations. Moses was instructed to take off his sandals, and to stop at a certain point. In that passage, we see a human encounter with sacred space.

What do we observe that human doing? He is navigating a stipulated freedom.

This is the foundational premise of this book. It is the assertion that in the various areas of possible biotech activity, we are in fact navigating a stipulated freedom. We are operating within God' permission, and with God's stipulations. Like Moses, we too must comply with certain instructions. Like Moses, we too must observe certain limits. Within those instructions and limits however, there is a freedom. Once we discover just how broad that freedom is, we will see how far humanity can travel through the realm of biotechnology. Our freedom, although stipulated, is an amazing bold permission of God. It is, in fact, a permission so bold, that it represents the biggest risk God could ever take.

What is the next step? It is to search out this act of permission that has given humanity its stipulated freedom. As one discovers this act of permission, one is also led to an understanding of what the stipulations entail. Interestingly, we find this act of permission in the very same place where we saw Moses navigating a stipulated freedom into sacred space. It is that same mysterious and smoky mountain, the chosen place for God to establish the boundaries of God's relationship with humanity. In the second chapter, let us return to that mountain in order to discover our stipulated freedom.

Chapter 1 Notes

[8] Exodus 3:5

[9] The manuscript of this interview can be found on the website http://www. PBS.org/faith and reason/ted-body.html

[10] The text of this speech can be found on the website http://www. stnews.org/news-1324.htm

[11] This verse is from the New International Version

[12] Many religious leaders have asserted that these Hebrew terms, and their corresponding pronouncements of judgment, are clear Biblical statements on homosexuality. Although this book is not about that subject, I think it is important to note that the Biblical word under discussion here, קָדֵשׁ (qadesh) is in reference to a very particular cultic role and does not reference homosexuality in a general sense.)

[13] Merriam Webster Dictionary, p.

[14] The Interpreter's Dictionary of the Bible, vol. 2, p. 817

[15] Eliade, Mircea. The Sacred and the Profane. *The Nature of Religion*, p. 26

[16] Childs, Brevard. Myth and Reality in the Old Testament, p. 78

[17] Clines, David J. *Sacred Space, Holy Places, and Suchlike*. In On the Way to the Postmodern: Old Testament Essays 1967-1998, Volume 2, p. 545

[18] Romans 13:5

[19] Buber, Martin. I and Thou. Translated by Walter Kaufmann, p.160-161

[20] Acts 17:28.

[21] Murphy, Nancey and Ellis, F.R. George. On the Moral Nature of the Universe. *Theology, Cosmology, and Ethics*, p.39

[22] Peacocke, Arthur. Theology for a Scientific Age. *Being and Becoming-Natural, Divine, and Human*, p. 29

[23] Kadavil, Mathai. The World as Sacrament. *Sacramentality of Creation from the Perspectives of Leonardo Boff, Alexander Schmemann, and Saint Ephrem*, p. 62

[24] Ibid, p.

[25] Heidegger, Martin. The Question Concerning Technology *and other Essays*, p.21

[26] Peacocke, Arthur. Theology for a Scientific Age. *Being and Becoming-Natural, Divine, and Human*, p. 31

[27] Alexander, Denis. Rebuilding the Matrix. *Science and Faith in the 21st Century*, p.239

[28] Kadavil, Mathai. The World as Sacrament. *Sacramentality of Creation from the Perspectives of Leonardo Boff, Alexander Schmemann, and Saint Ephrem*, p. 126-127

[29] Metzner, Ralph. Green Psychology. *Transforming Our Relationship to the Earth*.

[30] McNeill, John T (Ed). Calvin: Institutes of the Christian Religion. volume 1, p. 186

[31] Buber, Martin. I and Thou. Translated by Walter Kaufmann, p. 54

[32] Sachs, John R. The Christian Vision of Humanity. *Basic Christian Anthropology*, p. 17

[33] According to the International Society for Complexity, Information and Design Encyclopedia of Science and Philosophy, Neo Darwinism is defined as "the modern version of Darwinian evolutionary theory: the synthesis of Mendelian genetics and Darwinism. Darwin knew very little about the mechanism of variation; he merely recognized that whatever its source, phenotypic variation allowed for natural selection to operate. It was modern genetics that provided the key insight into the means by which variation in biology originated." (http://www.iscid.org/encyclopedia/Neo-Darwinism)

[34] Several independent studies failed to support Hamer's research. These included a study in 1998 by Dr. Allan Sanders of the NIH, and a study conducted by the University of Western Ontario in 1999.

[35] Hamer, Dean H. The God Gene. *How Faith is Hardwired into our Genes*, p.74

[36] http://www. pqasb.pqarchiver.com/washingtonpost/access/739293281.htm

[37] Lewis, C.S. The Abolition of Man. *How Education Develops Man's Sense of Morality*, p. 83

[38] Herzfeld, Noreen L. In Our Image: *Artificial Intelligence and the Human Spirit*, p. 6

[39] Eliade, Mircea. The Sacred and the Profane. *The Nature of Religion*, p.116

[40] Psalm 119:1

[41] Romans 1:20

[42] McNeill, John T (Ed). Calvin: Institutes of the Christian Religion, Book I, Chapter XIV, i

[43] For more on a synthesizing of religious ideas and metaphysics, see God and the New Physics , by Paul Davies.

[44] Kadavil, Mathai. The World as Sacrament. *Sacramentality of Creation from the Perspectives of Leonardo Boff, Alexander Schmemann, and Saint Ephrem*, p. 82

[45] Alexander Schmemann was a professor and later Dean of the St. Vladimir's Orthodox Theological Seminary in New York City. Most of his books were published posthumously.

[46] Moltman, Jurgen. The Church in the Power of the Spirit. *A Contribution to Messianic Ecclesiology*, p. 199ff

[47] Gilkey, Langdon. Nature, Reality, and the Sacred. *The Nexus of Science and Religion,* p. 204

[48] Peter Lombard (1100 – 1160) developed this guiding theology of sacraments in his tome the Book of Sentences. (Libri Quattuor Sententiarum). The sacraments are addressed in book IV.

[49] Philip Sherrard was an English author and scholar educated at Cambridge. Among the works for which he is best known is his collaboration in the complete translation of the *Philokalia*, a collection of texts by masters of the Eastern Orthodox tradition from the fourth to the fifteenth centuries.

[50] Philippou, A.J. (ED) The Orthodox Ethos: Essays in Honour of the Centenary of the Greek Orthodox Archdiocese of North and South America, p.134

[51] Ibid, p. 133

[52] Lossky, Vladimir. The Mystical Theology of the Eastern Church, p. 93

[53] Ibid, p. 240

[54] Peters, Ted. Playing God? *Genetic Determinism and Human Freedom*, p.14

[55] Cole-Turner, Ronald. The New Genetics. *Theology and the Genetic Revolution*, p. 45

[56] Peters, Ted. Playing God? *Genetic Determinism and Human Freedom*, p.26

[57] The discussion here about human "rights" is not fully developed. It certainly would be wrong to assert that the only way to curtail a sense of rights to manipulate nature is through a theocentric approach. In his book, After Virtue, Alasdair MacIntyre presents a convincing case that demonstrates the failure of the notion of "rights" in a non-theological perspective. The argument here is to differentiate the difference between seeing nature as fundamentally non-sacred compared to seeing that all of nature, albeit it having no intrinsic sacredness, is nonetheless sacred to God.

[58] http://www. en.wikipedia.org/wiki/Precautionary_principle.com

2

Discovering the Yellow Light

Introduction

It was Moses who, in the beginning of the first chapter, led the reader down the path of contemplating the restrictive nature of sacred spaces. It was learned that sacred space is space that has been set apart, and claimed as God's own space for God's purpose and will, and that the only way that any human can enter sacred space is by God's own invitation. Entering such space is not a human right, but rather a gracious permission that is initiated by God, and entered under God's stipulations.

It has also been seen, through the classic doctrine of General Revelation as well as the mainly Orthodox theology of the world as sacrament, that nature itself, consisting of all matter, space, time and energy, is indeed sacred to God. Nature has been chosen by God both to reveal God's self to humankind and to extend God's grace to humankind. Therefore, human manipulation of nature, which is brought to startling new levels in our sciences of biotechnology, is not a human right. If it can happen at all, it is by God's permission. In other words, the manipulation of nature is within God's exclusive realm of authority and power. Any permission that God would offer would be akin to God releasing some of that power to humankind.

Is there any evidence that God has released such power to humankind; If so, how much of that power? Have humans been given permission to exercise power over designer breeding of horses, but not of a human child? Do humans have God's permission to irrigate water but not to steer the direction of DNA? Undoubtedly, everyone must agree that humans have been given permission to exercise some powers over nature, but the controversies around biotechnology assert a certain "line in the sand", through which human powers are not allowed to cross. Some would articulate this prohibited territory as the part of nature that is sacred, while the rest of nature is not. This view would be rejected by the previous chapter's discussion of nature and the sacred. *All* nature is sacred to God. There is no line between sacred nature and profane nature. It would be an unfounded claim to assert that the chlorophyll molecule in a blade of grass is profane but the human gene CREB on Chromosome 2 is sacred. Both are matter. Both are a part of nature. As such, neither have an inherent sacredness unto themselves. Yet both are sacred to God.

Since all nature is sacred to God, it is only God who can give humanity a freedom to encroach into God's sole territory of power. Is there any evidence in Scripture that God has given such freedom to humanity? If so, does that evidence also show any stipulations that go along with that freedom? These are the questions that will be addressed in this second chapter. In this chapter, we will seek to create a bridge between an ancient piece of Biblical rhetoric and our modern science of biotechnology. As one searches for a deeper meaning of this Biblical concept, one discovers a dramatic freedom that God has granted to humanity. Along with that freedom, there is a strong stipulation. This stipulation, however, is not what one may think. It does not draw a line along the path of possible manipulations of nature beyond which humanity can not go. Rather, the stipulations that are found raise important ethical questions about every human manipulation of nature, from the grass chlorophyll molecule to the human genome. These stipulations are not so much about nature, as they are about the human community.

The issues of this chapter are critical for and foundational to the goal of creating a compass for the ethical questions raised by the various possibilities of biotechnology. Where does the important journey of this chapter begin? Interestingly enough, it is on that same mountain where we last left Moses during his encounter with the burning bush, and with sacred space. So let us return to Moses, as he stands in awe of that burning bush; For although Moses was not aware of it at the time, the journey of stipulated freedom was about to begin.

Returning to the Burning Bush

As one reads of the dramatic encounter between God and Moses as recorded in Exodus chapter 3, it must be kept in mind that this is the first time that Moses enters any kind of dialogue, or any kind of relationship, with God. As Moses negotiates his permission to enter the sacred space represented by the burning bush, he receives a commission from God that will require great courage and action on Moses' part. He is instructed by God to return to Egypt, the land from which Moses ran, and demand from Pharaoh the freedom of the Israelites, who have long been the slaves of the Egyptian rulers. With this frightening mission given by an unfamiliar God, it seems quite natural that Moses would want a better understanding of this God of the mountain. It is not surprising, then, to hear Moses pose his question to this God: *"If I tell the people that the God of their fathers has sent me to them, and they say to me 'What is his name?', what shall I say to them?"*[59] It seems like a simple and logical question: *God of this sacred space; what is your name?* A closer look at this question, however, reveals that something deeper is occurring in this dialogue. It sets the stage for understanding the yellow light of God's stipulated freedom for human manipulations of nature. Moses wanted to take the name of God back to Egypt. It is precisely this journey back to Egypt that will provide an insight into the deeper meaning of Moses' seemingly innocent request.

To Take a Name back in Egypt

Although it is clear that Moses is an Israelite, it is also known from his story in Exodus that Moses was a product of the Egyptian culture. In fact, the entire people known as the Israelites were very much a part of the larger cultures around them. As such, the religious faith and beliefs of the Israelites had some level of congruency and similarity with the cultures that were in close proximity, particularly the Egyptian culture. Of course, the defining hallmark of the Israelite faith is something that is in clear contrast to the surrounding cultures: their belief in one God, who was separate and distinct from the world. The religious cultures of the peoples who lived around the Israelites were mostly polytheistic and to some degree, pantheistic. The monotheistic faith of the Israelites marks a significant departure from the surrounding cultures, and has been often seen as the crowning contribution that the ancient Israelites have given to the modern world.[60] Noting that important difference, however, there are still themes of the Biblical story that find a stream of connection with the surrounding religious cultures. Research into the religious life of ancient Egypt has revealed that not only did the Egyptian people believe in multiple gods, but they also had multiple names for those gods. One recognized authority on ancient Egyptian religion, the German scholar Erik Hornung, states:

> Multiplicity of names is a fundamental feature of the gods which is common to all polytheistic religions. In particular, the great gods, such as Amun, Re, and Osiris, are not content with presenting themselves to their worshippers under only one name.[61]

The existence of multiple names of Egyptian gods was meant to protect them from the possibility that anyone, divine or human, would posses their true (and therefore hidden) name. The reason for this was the basic belief of what happens when one possesses the true name of another. Such possession of a name would result in the possession of the very essence, and power, of the named individual. The priestly magicians of ancient Egypt, through their cultic rituals and practices, demonstrated this correlation between having the name of a god, and therefore possessing that god's power. Adolph Erman, another German authority on ancient Egypt, described in his book <u>Life in Ancient Egypt</u> how the court magicians relied on the possession of the names of the gods. Erman writes:

> It was of course especially effective if instead of using the usual name of the god, the magician could name his *real name*, that special name possessed by each god and each genius in which his power resided. He who knew this name, possessed the power of him who bore it.[62]

Erman goes on to elaborate on one of the primal stories of ancient Egyptian religion, namely the story of the sun god Re. The story well illustrates the Egyptian belief in the possession of a name:

In the primeval ages of the world, the sun god *Re* appeared on the dark ocean of the god *Nun*, and undertook the government of the world. This did not happen without a struggle; finally however the victory remained with *Re*, and the "children of the rebels" were delivered up to him on the terrace of the town Chmunu. He now reigned in peace as "King of men and gods," and as long as he was in full possession of his powers, no one attached his government. But his youth was not eternal; his limbs became stiff with old age, his bones changed to silver, his flesh to gold, his hair to real lapis-lazuli. Then happened what happens to all earthly kings as they grow old: his subjects became rebellious, more especially the wise goddess *Isis*, who was wiser than all men, than all gods and spirits. She new all things in heaven and earth as well as *Re* himself, but there was one thing she did not know-and this want of knowledge impaired her power-the secret name of *Re*. For this "god of many names" kept his special name secret, the name upon which his power was founded, the name which bestowed magical might on those who knew it. [63]

According to this ancient Egyptian myth, Isis found a way to get Re to reveal his true and hidden name. She conjured up a poisonous snake, which proceeds to bite Re and inject its excruciating toxin. Although Re struggles to protect his true and hidden name, the pain becomes too great, and he finally releases his true name to Isis. In his book Conceptions of God in Ancient Egypt, Erik Hornung writes:

> The story, which was intended as an effective spell against snake bites, shows clearly that the chief god, at least, possesses secret names in addition to those he bears in cult and myth. Some texts give the impression that there was a ritual prohibition against pronouncing some divine names; Ramesses IV asserts on his stela from Abydos that "I have not pronounced the name of Tatenen." It is unlikely that the name referred to is the well known name of Tatenen itself; it must rather be another name which is to be kept secret. [64]

In ancient Egypt, it is clear that the possession of the name of a deity equates to a possession of that deity's power. To take a name, then, means to harness the power of the named individual. When one compares the ways in which the names of deities are sought by others and protected by the deities in the religious milieu of ancient Egypt with the ways in which humans grapple with Yahweh in the biblical texts of the Hebrew Scriptures, a bridge of similarity begins to emerge.

To Take a Name in the Hebrew Scriptures

In Exodus 3, Moses wants to possess God's name. In the Hebrew Scriptures, he is not the first or only one to do so. In Genesis 32, Jacob seems to be asking for the name of God. His question only elicits a question in return, "why is it that you ask after my name?"[65] In Judges 13, Manoah is in dialogue with the mysterious 'Angel of the Lord' concerning the promise of a child (Samson). When Manoah requests the name of this important visitor, the response seems so much in character with the beliefs of ancient Egypt:

> *And the angel of the Lord said unto him, why do you ask after my name, seeing that it is secret?*[66]

The repeated pattern of a human seeking possession of God's name with God's hesitancy to reveal it indicates that the ancient Israelites, from whom our Judeo Christian Scriptures emerged, shared a similar understanding of the power of the Deity's name with the belief system that has been observed in the ancient Egyptians. As we leave Egypt behind, we see that this core understanding of the power of a name, particularly the name of God, carries forward into the developing faith of the Israelites. In the <u>Anchor Bible Commentary Series on Exodus</u>, William Propp comments on the obvious interest that the Hebrew Scriptures have for the name of God. He writes:

> Why such fascination with deities' names? The noun *sem,* can also mean "self" or "essence", and scholars often observe that to know an object's name is, in the world of magic, to possess power over it. This is the likely import of Adam naming woman and the beasts (Gen 2:19; 3:20), or of Yahweh renaming the Patriarchs (Gen 17:5, 15; 32:29; 35:10). But what about a Deity's name? In ancient Egypt, knowing the gods' secret names gave humans a degree of

mastery over them. Similarly, in later magical folklore both Jewish and Gentile, God possesses a secret name (not "Yahweh"), the knowledge of which confers some of his power upon humans; it was supposedly engraved, for example, on Moses' staff. Humans in the Bible, then, are understandably eager to learn the names of deities, and the latter are understandably chary of disclosing them.[67]

In the Hebrew Scriptures, then, one can see a correlation with the Egyptian belief in the power contained in a name. The characters of the Bible want to possess God's name, because in doing so they will be able to harness the power of God. God is rather reluctant to share his name, because doing so would be an act of *permission*. By releasing his name to humanity, God would be granting humanity permission to harness God's power. When God refuses to do so, God is in essence claiming something to be for God's own use and therefore off limits to humanity. This comes back full circle to the concepts established in the first chapter. If God's power is set apart by God for God's exclusive use, then God's power is sacred space. God's name is sacred space.

In the Hebrew Scriptures, it is clear that the power of God is witnessed within the context of nature. This power of God is a creative power. It is the power that enables babies to be born, and seas to part. It is the power that makes mountains rise, and tempests churn. It is not a very far leap to go from the statement *the power of God* to the statement *the power of God over nature*. Although our Judeo Christian faith certainly asserts that God's power exists beyond nature, we are much more apt to visualize God's power in relationship to nature.

This, then, brings us to a point of summary before we continue. In the Hebrew Scriptures, we find a belief akin to the Egyptians. The possession of God's name indicates the freedom to harness and use the powers of God. These powers of God are observed within the realm of nature, in which God creates and manipulates. They are the very powers that work within nature. If God has set nature apart for God's own purpose (i.e. made nature sacred) then nature is the realm of God's exclusive power. If God were to release any of his power to humanity, it would be paramount to God inviting humanity into sacred space. Although there may not be an obvious avenue in which God extends his power to humanity, there is a symbolic one. It is the hidden, secret name of God. If that name is strictly protected, sacred spaces are off limits. The light is red. If humans are successful (as tried Jacob and Manoah) in tricking God to release his name, then the humans have wrestled the powers out of God's hands. The light to proceed in the full domination of nature is green. If, however, God willingly reveals his name, or at least some attributes of it, a yellow light begins to emerge. God is inviting humanity into the realm of God's power over nature. Since it is permission rather than a human right, God can stipulate the freedom offered in any way that God chooses.

As Moses stood before the burning bush, having just received a frightening mission from God, he had asked his question: *When I come to the children of Israel, and shall say to them 'the God of your fathers has sent me to you', and they say to me 'what is his name?', what shall I say to them?* God responded by giving that familiar, and downright perplexing, response: I AM THAT I AM.

Throughout the years of Biblical scholarship, learned people have disagreed with each other over the meaning of this answer. Some believe that God was, in fact, revealing his name, albeit in a less than straightforward way. These linguistic contortions of the Hebrew word *Yahweh* are often dismissed as too contrived.[68] A more widely accepted interpretation of this answer is that it conveyed the concept that God was too big to be contained adequately in words. In the New Century Bible Commentary, we read:

> The various translations are all obscure, but they are an attempt to express the independence, the self-sufficiency, the self consistency or the eternity of God, and the idea that the nature and attributes of God transcend thought and language.[69]

Perhaps the strongest advocate for this position was the respected Biblical authority S.R. Driver, who wrote his commentary on Exodus in 1918. Driver described this phrase, I AM THAT I AM, as an *idem per idem*, a circular form of rhetoric that indicates an inability to describe further. A modern example of this form of rhetoric would be the phrase "*What will be will be.*" In this form of rhetoric, there is no attempt to define something that seems indefinable. Driver describes the *idem per idem* this way:

For the form of sentence called the *idem per idem* construction, which is idiomatic in both Hebrew and Arabic, where the means, or the desire, to be more explicit does not exist. The second idem in the sentence is a simple future: It must not be emphasized as though it meant "wish to".[70]

. If this is the case, then, God does not give Moses a name, but not because God doesn't wish to reveal it. He doesn't give a name due to the fact that human language is unable to contain the essence of God's nature. The *idem per idem* is used because the means to be more explicit in human language do not exist.

Still others, however, offer a different interpretation. One such example was written by Jack R. Lunbom in his article *God's Use of the Idem per Idem to Terminate Debate* that appeared in the Harvard Theological Review. In this article, Lunbom offers by way of example many cases in both Biblical and modern speech in which this rhetoric tool is employed not because there were no words that could be used, but simply because the speaker wanted to terminate the discussion. In drawing the line of comparison between these examples and the story of Moses at the burning bush, Lunbom concludes that the purpose of God's response was to end debate. God wasn't going to give Moses his name. In concluding his article, Lunbom well articulates the challenge of this passage:

The rhetoric of a passage is then the key to meaning and interpretation. In Exodus 3 scholars continue to debate over whether God gives or does not give Moses a name. Obviously, there is tension in the text, but must we come down on either one side of this question or the other? Why not rather acknowledge the tension and simply leave it at that? When the *idem per idem* terminates debate there is always tension because the answer it gives will be perceived at the same time as a non-answer.[71]

The best one can say, then, is that the case of Exodus 3 is ambiguous. Even so, it does provide valuable insights into an important theme that is so foundational to the understanding of a stipulated freedom to harness God's powers over nature. If nature is sacred to God, only God has the right to exercise power over it. God's power is symbolically portrayed in the Hebrew Scriptures by God's *name.* Humans, in an effort to harness that power, tried to ascertain that name. In his desire to continue the prohibition of human possession of his powers, and therefore human encroachment into sacred space, God refused to release his name. As Manoah learned, it is a secret. Even Moses, as he received his commission from God to return to Egypt and lead God's people out of Egyptian slavery, was denied access to God's name.

That, however, is not the end of the story. In many ways, it is only the beginning. Indeed, God did instruct Moses to free God's people from their bondage in Egypt. Once freed, however, they were to return to that mountain. Once they returned, God was ready to establish a new framework in which he was going to relate to his people. In this new framework, God was going to take an amazing risk. At this point of God's relationship with his people, God was going to release his name to humanity. Our stipulated freedom was about to emerge, along with some clear directions of how we were to navigate through it. Major responsibility, and significant risk, is contained in this stipulated freedom. As Propp states:

> Unlike Jacob and Manoah, Moses ultimately succeeds in wheedling the divine name out of Yahweh. Thereby he transmits to Israel and humanity a mighty trust, dangerous to misuse. The spreading knowledge of Yahweh's name is a major theme of Exodus, and of the Hebrew Bible as a whole.[72] (Propp, 224)

Our journey therefore must quickly acknowledge the return of Moses to Egypt, the plagues that finally resulted in Pharaoh's release of the Israelites, the defining moment at the Red Sea, and the encampment of the Israelites around the very same mountain upon which Moses met the burning bush. Now that God had gathered his nation, it was time to establish the rules of the relationship between God and his people. A core component to that relationship, and to the entire religious law and practice of the Jewish people, was the *Ten Commandments.* The commandments are seen as a core instruction for human behavior, not only in Judaism, but also in Christianity, Islam, and even non religious society. Within the Ten Commandments, we return to this theme of God's name, and to the discovery of the stipulated freedom that begin to show us the way for the future road of biotechnology.

The Ten Commandments
and Our Stipulated Freedom

As we arrive now at the Decalogue, we come to an all important point in the total development of this book. The assertion that will be made is that within the Ten Commandments, some basic questions are answered: Does God allow humanity to manipulate nature? If so, are there any stipulations that accompany that permission? Are there any limits? Once we observe answers to these questions, we will discover a roadmap to guide us as we return to the characters of Greek mythology, who serve us as symbols of some of the most controversial aspects of modern biotechnology.

Perhaps the initial question that one would ask is: Why the Ten Commandments? Is this not simply one text among many from the Scriptures that could be used to create ethical evaluations of biotech issues? If the Decalogue is equal in weight and importance to all other Scripture, then choosing it as a basis for ethics is rather arbitrary and subjective.

To the contrary, it is the intent of this book to assert that the choice of the Decalogue as a foundational basis for doing ethics is not arbitrary, and that the Decalogue itself stands out as a unique and authoritative passage. To assist in this assertion one can call upon the now Emeritus Professor of Old Testament Theology at Princeton, Dr. Patrick Miller. In an essay entitled *Divine Command and Beyond: The Ethics of the Commandments*, Miller writes:

> Yet one cannot be too casual at this point. The Decalogue is not just any text; it is a foundational text, if there is such a thing. Only here in the Scriptures does God speak to the whole assembly of God's people. Therefore, one can hardly say that the character of such speaking as largely command is of little importance or simply one option among others that could have provided a moral grounding and framework for the community of faith that began with the Israelite people. No, the divine command is the character of the divine word at a critical point. When God speaks to the community, it is in the form of command, and that form assumes fundamental obligation.[73]

Having said that, however, one must acknowledge that any attempt to create an ethical bridge from the Decalogue to modern society will itself foster a great deal of concern and controversy. In beginning his book on the Ten Commandments and Christian Ethics, Jan Milic Lochman states:

> Anyone seeking to develop an outline of Christian ethics today on the basis of the Ten Commandments must be quite clear as to the controversial character of such an undertaking. It can be challenged not only by pointing out how vital it is for an ethic to speak to the contemporary world but also on fundamental theological grounds.[74]

Lochman goes on to articulate the basic reasons for this. The Ten Commandments, the argument says, are extremely ancient, and as such offer little insight into modern life. In addition, they grow out of a particular historical context of time and place, and assume relationships (such as between men and women) that are much different today. Many other reasons have been postulated why the Decalogue should not be seen as a timeless code of morality and ethics.

Lochman states that the Ten Commandments can not be *rigidly* applied to the ethics of modern life. If one tries to use these precepts as prescriptive rules and regulations for modern life, one will quickly enter the realm of legalism. Lochman writes:

> If laws are interpreted legalistically, they shackle ethics to the conditions which prevailed at the time of their promulgation. They become backward looking and fail to speak to the conditions of our contemporaries in their (and our) own very different social context and stage of cultural development.[75]

The application of the Ten Commandments as a legalistic rulebook for modern life is indeed problematic, and it is certainly not the proposition of this book. There are some Biblical scholars who make a convincing point that such an application of the commandments was an unfortunate and unintended aspect of the religious life of the Israelites as well. Johann Jakob Stramm, a leading authority on the Decalogue, presents this case nicely in his book <u>The Ten</u>

Commandments in Recent Research. Another renowned Biblical scholar, Gerhard Von Rad, presents this argument in his acclaimed Old Testament Theology (Vol. 1.). Von Rad illustrates that the original intent of the Ten Commandments was far different than what had emerged over time. How did the original understanding get lost? Von Rad writes:

> The way leading up to the end of this understanding of the law was opened up as early as the post-exilic period. It is, of course, a matter of a long and partly unseen process. The end was reached at the point where the law became an absolute quantity, that is, when it ceased to be understood as the saving ordinance of…the community of Israel…In this way it finally became a "law" in the normal sense of the term, a law which had to be adhered to word by word, indeed letter by letter. [76]

Von Rad believes that the book of Deuteronomy, with its spirit of internalizing the law of God into the themes of worship and love, was a concerted defense against the rising tide of legalism that was sweeping over the Decalogue. Von Rad states that "in concerning itself so earnestly with the inner, the spiritual, meaning of the commandments, Deuteronomy rather looks like a last stand against the beginning of a legislation."[77]

If the Ten Commandments are not meant to be seen as law, then what are they? The first step in discovering the answer to this question is to recognize that these commandments are given within a larger context, in which the Israelites have gathered under a new sense of identity. This new sense of identity was two fold. First, they saw themselves as a "people", rather than different tribes and clans. This emerging sense of a human community was coupled with the second element. They saw their human community in a new relationship with God. God had saved them from bondage in Egypt, and from certain death at the hands of the approaching Egyptian army. God had gathered this new nation together, and was promising them a new land. A new, wonderful thing was happening. In his book, The Sinai Myth, Andrew M. Greeley imagines the scene:

> What are we to make of this whole affair? A group of Semitic slaves escapes from Egypt. Some of their neighbors have a

peculiar experience near a sacred mountain in the desert. They come together with other tribes around a desert oasis. There is no political unity and no one strong enough to become king. Out of sheer necessity of maintaining some sort of peace with one another an amphictyony emerges, that is, a tribal confederation based on a common religious belief. The tribes discover that this religious belief, centered on one God, isolates them from neighboring cultures and forges them into one people.[78]

Seen in this way, it becomes clear that the Ten Commandments were not originally received as elements of law and restriction. Rather, they are seen as a part of an amazing salvation moment. It is a saving moment in which a *covenant* is being established between God and his people. Von Rad writes:

> the most important question is that of the proper theological evaluation of the commandments. New force has now been given to the idea that Israel understood the revelation of the commandments as a saving event of the first rank, and celebrated it as such. In all circumstances the close connection between commandments and covenant must be kept in view.[79] (Von Rad, p. 193)

We have seen, therefore, that the Decalogue was given within the context of a celebration of salvation, and that a crowning moment of that celebration was the presentation of a covenant. It has been well proven in Biblical studies that the events recorded in Exodus resemble the typical treaties and covenants of the time period. One famous comparison has been made with the ancient Hittite treaty. Stamm offers a good summary of this:

> That the nature of the Israelite covenant festival is connected in some way with the Hittite treaty formula can scarcely be contested. As a reminder, we append once again the following four parts of the Israelite festival: 1. Paranetic prologue; 2. Proclamation of the commandments; 3. Making of the covenant; 4. Blessing and curse. The elements from the Hittite treaties which correspond to these are: 1. Preamble

and historical prologue; 2. Conditions of the treaty; 3. Conclusions with blessing and curse.[80]

Comparing the covenant that God makes with the Israelites in Exodus with the Hittite treaty formula offers important insights, not mainly by the similarities, but even more so, by the differences. For whereas other treaties do look rather rigid and legalistic, the covenant that God makes with the Israelites is rather broad and unspecific. In making this observation, Greeley states:

> If it were simply a treaty binding together a number of tribes under a new leader, the covenant would have been with Moses. He was the leader of the tribes. But Moses did not present himself as the overlord with whom the covenant was made. On the contrary, he was the go-between, humble, frightened, and not always reliable. It was no earthly leader Israel dealt with, and it was no brilliant monarch who forged them as a people. It was Yahweh, the Lord of creation. The stipulations he makes are totally different from those of other suzerains. He is not concerned about military might and pledges of assistance in time of war. He does not need these things; he demands the fidelity of his people, a fidelity evidenced by their loyalty to him. Nor are the stipulations he imposes on his people complicated or rigorous, not, at least, in their most primitive form.[81] (Greeley, p. 48)

The stipulations are not complicated or rigorous, he states, "in their most primitive form". Unfortunately, the processes of legalism would soon devour these basic commandments, and their original sense was lost. Von Rad, in his beliefs about the role of Deuteronomy in seeking to stand off the tide of legalism, finds an echo in Greeley:

> In later years, these stipulations would be expanded into a complex legal system, imposing a vast number of obligations, which purported to measure the amount of response to Yahweh's commitment. There was considerable resistance to this development. Deuteronomy may well have been the last effort to resist the rigid formulation of obligations.[82]

Resisting the tendency to rush to the legalistic interpretations of the Decalogue, one is left to contemplate their basic and most primitive meaning. If one strips away all of the previous conceptions of the Decalogue, what is found is something that is surprising, and perhaps startling. The meaning of this covenant, and of these commandments, is presented in the very first words of God: The preamble to the Decalogue. God said…

> *I am the Lord thy God, which have brought you out of the land of Egypt, out of the land of bondage.* (Exodus 20:1)

There are two remarkable things about this preamble. The first is this: *It establishes the context of this relationship between God and people as a relationship of freedom.* This covenant, and this relationship, is born out of freedom. It is important to realize that freedom is the overarching theme of this covenant, and of the Decalogue itself. Jan Milic Lochman implores us not to lose sight of the importance of this preamble of freedom:

> The 'signature', the 'sign outside the bracket', affects everything inside the bracket. It indicates the direction for the whole of the Decalogue and for each of its individual components, and also, in a broader and more inclusive sense, for the whole field of *theological ethics*.[83]

The basis of the Decalogue is freedom. It is not law. Walter Harrelson, author of <u>The Ten Commandments and Human Rights,</u> states:

> It would be wrong to treat the Decalogue as law, to see it as some onerous burden laid upon the people by a stern and righteous judge. The Ten Commandments are rather to be seen as Israel's great charter of freedom.[84]

It takes some adjustment of thinking to see the Decalogue as primarily a statement of human freedom, especially since most of the commandments begin with the words *"Thou shalt not"*. To see the relationship that God extends to humanity as one of basic freedom, however, begins to shed light on why the commandments seem to be negative and prohibitive. As we have already seen, Gerhard Von Rad opposes the idea that the Decalogue was given as law, in the primary understanding of the term. He writes:

> for a "law" in the narrower sense of the word, instructions for the moral life, the Decalogue lacks what is of first importance-the positive filling out-without which a law is scarcely conceivable. Instead, apart from the two well known exceptions, it refrains from any attempt to set up positive norms for the affairs of life. It confines itself to a few basic negations; that is, it is content with, as it were, signposts on the margins of a wide sphere of life to which he who belongs to Yahweh has to give heed.[85]

Von Rad's image of the commandments as "signposts on the margins of a wide sphere of life" seems to be fitting language for a road navigation such as the one of this book. Perhaps one can better understand, now, what that phrase means. If the Sinai Covenant is seen as first and foremost a moment of human liberation, in which God declares us free, then freedom is our course. Freedom is our environment. Freedom is the journey of life that God grants to his people. Once this sense of freedom is firmly established, then (and only then) does the covenant move toward the statements of "*thou shalt not*". These statements are not legalistic rules and regulations that are meant to enslave us. Rather, they are merely the "signposts on the margins of a wide sphere of life" (Von Rad). The commandments are the *stipulations* of our freedom. A stipulation is a condition that is attached to a contract or covenant. It is not the main contract, but rather an attached agreement of how the parties of the contract will conduct themselves. In the covenant that God makes with his people in Exodus, the main contract is the act of salvation and liberation that has resulted in freedom for the people. The commandments are the stipulations that are meant to keep that freedom within certain boundaries of God's choosing.

This is a good point to insert a brief word about the continuation of the Decalogue in the New Testament Scriptures. In his own ministry, Jesus demonstrated a resistance to seeing the Ten Commandments in the legalistic way that so dominated his day. When he was asked to speak on the Decalogue, he summed it up in this way:

'Love the Lord your God with all your heart and with all your soul and with all your mind.' This is the first and greatest commandment. And the second is like it: 'Love your neighbor as yourself.' All the Law and the Prophets hang on these two commandments."[86]

As will be seen later in this book, the theme of freedom was foundational to early Christian theology. Here it is important to note that although Jesus and the early Christian writers may be seen as breaking with the law and the commandments, it may actually be said that they were returning to the original intent and purpose of the Decalogue.

As we have returned to the desert mountain with Moses, we have stopped to investigate the important features of the Sinai covenant, and of the Ten Commandments. We have discovered that in both ancient and modern times, efforts have been made to legalize these commandments and apply them as rigid prescriptions of morality and ethics. Stripping the Decalogue of this human tendency, we have discovered something important. The covenant of God is first and foremost a covenant of freedom, and the commandments themselves offer stipulations that serve as guideposts. Since the basic state is freedom, and the only limiting factor is the stipulation, it can be concluded that anything not stipulated is within the realm of human freedom. That is no small feature, especially when we turn our attention to the main issue that has brought us back to this mountain. It is the name of God, which serves as a metaphor for the power of God over nature. Where does this fit in the startling freedoms of the Sinai covenant? That question brings us to the second bold feature that is discovered in the preamble to the Decalogue.

The Name of God and the Ten Commandments

In the preamble of the Decalogue, God does two amazing things. We have already seen one of them. God declared his people free. One might automatically gravitate to the obvious freedom represented here, that of the release of bondage from their Egyptian slave masters. There is, however, an additional-and much larger-freedom implied in these opening words. God says: *I am the Lord your God*. It is a startling self disclosure, which should not be underestimated. The actual literal translation of this phrase is *"I am Yahweh, the God of you.*"[87] It would be misrepresenting the case to assert that this was the first time God's name was disclosed in the Exodus narrative. The reality is that the opening phrase of the Decalogue had been used several times earlier in the narrative, as the New Interpreter's Bible Commentary states:

> Thus God speaks the same powerful formula that has been reiterated throughout the Exodus narrative (cf. 7:15), in which the formula is designed to reassure Israel and to challenge Pharaoh.[88]

Even so, however, there seems to be a qualitative difference here. The name of God is purposely given as a first act of a new life of freedom for the people of God. God's disclosure of God's own name becomes the foundation upon which the entire covenant will be built. As Biblical scholar John Durham wrote:

> Far more is being declared here than any treaty ever claimed, above all in Yahweh's self-revelation and self-giving. As Muilenburg put it, these 'first words' of Yahweh to Israel, 'indispensably prior to all that is to follow', are 'the center and focus of the whole Pentateuch' and 'the very heart of the whole Old Testament.'[89]

Given our earlier discussions on the import of meaning associated with having the name of a Deity, it is time to make two key observations from the preamble of the Decalogue. These two observations emerge out of the two basic word constructions of this passage. The first observation grows out of the first part, *I am Yahweh.*

One should keep in mind that the religious context of this emerging faith held to the idea that to possess the name of a deity meant that one could possess, harness, and use the powers of that deity. This idea may quickly be dismissed as primitive superstition, and one could build a case that shows that the Hebrew Scriptures saw it as superstition as well. Whether this belief has validity, or is empty superstition, is really of no consequence to our interests in it. One should seriously doubt that God had any fear that someone might discover his name and harness his powers, much like when one finds a genie in a bottle and has control over him. The literal application of the belief in the power of names is not important. What is important is that the Bible understood this belief in the power of names, and could use that belief as a symbol of purpose and meaning. The actual disclosure of the name *Yahweh* is less important than the fact that God disclosed it. Understanding the symbolic meaning of a deity's name, God was making a bold statement: "You may have my name"; "You may share my powers"; "You are allowed to enter sacred space".

The covenant of God with Israel begins with a sweeping and far-reaching human freedom. This freedom includes the possession of God's name, which symbolically invites us to share in God's powers. We have already seen that God's powers, or at least our understanding of them, operate within the realm of nature. Therefore, we can return to our basic questions of this book with a developing answer. Is nature sacred? We have determined that it is not sacred in and of itself, but is made sacred by God. May humans harness nature, exercise creative powers over it, manipulate it and change it? Since nature is sacred to God, we do not have the *right* to do these things, but could do them if God granted us the permission to do so. Does our faith tradition offer evidence that God has given such permission? In what many believe to be the most basic and foundational document of our Judeo-Christian faith, the Sinai Covenant and its Decalogue, God gave us his name, the symbolic permission to take hold of God's powers. From a theological perspective, then, we can positively state a foundational freedom for the human manipulation of nature. This freedom to manipulate nature is not dependent on a narrow idea of what nature is. From metaphysics to nanotechnology, from the farthest planet we can reach to the microscopic world of DNA, it is all nature. If God has given us the freedom to exercise power over nature, then we should not shrink back from the edge of possibilities, as if we are approaching sacred space. If all of nature is sacred space to God, we are already operating within it. The lines we draw between mundane space and sacred space are artificial, and reflect more our discomfort with a broadened horizon of nature than with the sacredness of nature itself. In President Bush's 2006 State of the Union Address, he offered this brief paragraph:

> A hopeful society has institutions of science and medicine that do not cut ethical corners, and that recognize the matchless value of every life. Tonight I ask you to pass

legislation to prohibit the most egregious abuses of medical research: human cloning in all its forms, creating or implanting embryos for experiments, creating human-animal hybrids, and buying, selling, or patenting human embryos. Human life is a gift from our Creator and that gift should never be discarded, devalued or put up for sale.[90]

How much of President Bush's resistance of biotechnology is on legitimate ethical ground, and how much is an artificially imposed line that he set due to his own discomfort with widening horizons of possibility? It is so difficult to let go of the imaginary lines between mundane nature and sacred nature. We resist the notion of a radical freedom that God has given. Having said that, however, is it fair and right to boldly claim no limits to this human freedom? This question brings us to the second part of the preamble to the Decalogue. God said: I am Yahweh, *the God of you.*

Yahweh, the God of You

The gift of God's name, and its corresponding permission to share in God's creative power, is offered in the context of a relationship. It is the relationship between God and God's people. Understood in terms of this relationship, the freedom that is given must be received and used within the boundaries of that relationship. The freedom is genuine, and quite expansive, but it is not limitless. It is limited by the relationship. The freedoms offered must not offend that relationship. The freedoms must be exercised within the general scope of the values of the one who extends this relationship and this freedom, namely God.

It is understandable, therefore, than once freedoms have been offered (*I am Yahweh*) and a relationship has been established (*the God of you*), the author of this covenant would proceed to establish certain stipulations. These stipulations are meant to keep the broad exercise of granted freedoms within the boundaries of the relationship. To this point, we once again recall the words of Von Rad when he said of the Decalogue: "It confines itself to a few basic negations; that is, it is content with, as it were, signposts on the margins of a wide sphere of life to which he who belongs to Yahweh has to give heed."[91] Lochman echoes these thoughts in his book Signposts to Freedom:

If there can be no question of abandoning the Ten Commandments to the legalists, neither can we leave the concern for freedom, written as it is into the very fabric of the Decalogue, to the libertarians. Genuine human freedom is not just a blank cheque for us to fill out as we please. Precisely here, in this creative dialect of the history of freedom documented and disclosed in the Bible, I find the permanent and positive human significance of the Ten Commandments.[92]

With this, then, we come to one of the commandments, one which offers a negation on the use of God's name. It is important that one view the progress to this commandment in the right order. First, there is *liberation*. God has freed his people. Freedom reigns supreme. Second, there is *permission*. God revealed his name. God offered a symbolic gesture that granted permission for the human sharing of God's powers over nature. Third, there is *relationship*. Humanity does not live unto itself. We are in relationship to God. We are God's partners. As Ted Peters claims, we are God's "co-creators."[93] Fourth, then, there is *stipulation*. In one of the commandments (third by most listings) God said: *You shall not take my name in vain.*[94]

It is important to view this stipulation in its relationship to the preceding points to avoid an obvious and common pitfall. In ancient times as well as modern, people have gravitated to the literal and the prohibitive meanings of this commandment. For many, it is one of the great THOU SHALT NOT's...and the whole amazing point of freedom is completely missed. For ancient people, it was interpreted quite literally, to the point that the actual Hebrew name יְהֹוָה (YHWH) was not pronounced, and even to this day we are not certain of the name's correct spelling or pronunciation. As the Hebrew Scriptures were translated into the Greek, the name Yahweh was replaced with the Greek word κυριου, which leads to our English translation "Lord."[95]

While it is true that there is a definite negation contained in this commandment, one must be sure to read the negation in relation to the implied affirmation, just as one must always understand a stipulation in relation to a contract or covenant. The covenant is a covenant of freedom. The implied affirmation contained in the words *"you shall not take the name of God in vain"* is *"you can take the name of God in every way except...*" There is, here, a bold permission, and an extreme freedom. This freedom is stipulated by one single "signpost": You shall not take the name of God *in vain*. We are given the permission to harness the powers of nature, so long as we do not do so in vain. In his book <u>The Ten Commandments and Human Rights</u>, Walter Harrelson states:

> The third commandment carries out the theme of Yahweh's exclusive claims upon his people. Yahweh's power is available for the maintenance of the world and for the care and protection of his people...But there is a limit beyond which the community and its individual members dare not go in the claiming and use of God's power. That limit is the subject of the third commandment.[96]

We come, then, to the pivotal issue of the stipulation. It is important to note, especially in our political environment of creating lines in nature-separating mundane nature from "sacred nature"- that the stipulation makes no distinction in types of nature. It does not say "you may exercise power over the genome of wheat but not the genome of humans." It does distinguish between fertility drugs and cloning. To the contrary, in this case nature is nature. The freedom applies to all of nature. Likewise, the stipulation applies to all of nature. God has given us the freedom to participate in God's power over all of nature, so long as we do not exercise that power *in vain*. Our job, then, is to operate within the realm of our God given freedom, without crossing the boundaries of our stipulation. As our title indicates, we are *navigating through a stipulated freedom*. In order to safely navigate this course, one must gain a better understanding of this third commandment, and what the words "in vain" might mean.

Taking a Name *in Vain*

Any exegetical study of this third commandment will quickly show that there is a great deal of ambiguity in terms of its interpretation, as can easily be seen in the major Bible translations available today. In the NIV version, Exodus 20:7 reads "You shall not misuse the name of the LORD your God, for the LORD will not hold anyone guiltless who misuses his name." Many other versions, including the King

James Version (both old and new), the American Standard Version, and the Revised Standard Version (also both old and new) offer similar translations. In other versions, however, a different interpretation is offered, which translates the meaning into a much more specific prohibition of making a false oath upon the name of God. How is it that many versions agree that the meaning is a more general "take the name of" or "use the name of", while other (and mostly older) versions apply a much more specific meaning, that of "swearing an oath by the name of"?

The reason for this discrepancy is the ambiguous meaning of the Hebrew word נָשָׂא. It literally means "to lift up", and in this case would be literally translated "to lift up the name of God". In other Old Testament passages, such as several verses in Ezekiel chapter 20, this word is used to literally describe the lifting of a hand in making an oath, which is often interpreted with the single word "swore", such as in an oath. In Exodus 20:7, the word נָשָׂא is used, but without the language referring to one's hand. This has led some Bible translators to assume that the meaning of an oath is implied. This view is represented by Herbert Huffman, who states:

> With reference to the specific Biblical evidence, the initial phrase of the commandment, literally "You must not lift up the name of the Lord your God frivolously/falsely" is likely elliptical for the more expanded form "You must not lift up (your hand and speak) the name of the Lord your God falsely/frivolously. "Lifting up your hands" is a well established phrase referring to the gesture of swearing.[97]

The problem with this interpretation is that it is based on something that is believed to be implied, and implications are not as solid as observable facts. A second problem with this interpretation is that there is another Hebrew word that means "to swear an oath", which is the word שָׁקַר. This happens to be the chosen word for one of the other commandments that specifically addresses oaths, as found in Exodus 20:16: "*You shall not give false testimony against your neighbor.*" It would be highly unlikely that two different words would be chosen to express the same central concept within the same text. For these reasons, the more general translation of "take the name of God" is preferred to the more specific translation of "swear an oath by the name of God", and we certainly have the weight of many modern translations to support this claim.

The commandment, then, speaks of "taking", or "using" the name of God, which can be connected with the age old belief of possessing and utilizing the powers of a deity. It is not a total prohibition of taking the name of God. In fact, it leaves tremendous room for human freedom. It provides but one "signpost"; one stipulation of this human freedom. We can not take the name, we can not possess God's powers over nature, *in vain.*

The Hebrew word chosen here is the word לַשָּׁוְא. It is often translated as "lightly", or "with idleness". These words would indicate a use of power that does not carry the *importance* of that power, or does not express the *purpose* of that power. In this sense, the forbidden use of God's name is not merely an evil or destructive use. If that were the case, this would be a rather easy commandment to follow. Evil and destructiveness are usually easy to spot, and people who wish to navigate an ethical course would have some very clear markers. There are some who do see a more concrete sense of evil in this Hebrew word. One example is Walter Harrelson, who writes:

> The Hebrew expression *lassaw* has often been translated "in vain", with the meaning to treat Yahweh's name lightly, to use it idly, to treat it with light contempt. I do not find any occurrences clearly supporting such a translation. The word is stronger than such a translation indicates. It is used in parallelism, or near parallelism, with *to 'evah* in Isaiah 1:13: 'Bring no more vain offerings; incense is an abomination to me.' A much better translation would be 'bring no more destructive offerings' or 'offensive offerings'. Rather than being an expression for emptiness or insubstantiality, the term carries with it active power for harm.[98]

Although evil and destructiveness would certainly qualify under the banner "in vain", one would be doing the concept an injustice if one narrowed its application to this. Many of us learned in catechism that there are two types of sin: sins of *commission* and sins of *omission.* Sins of commission are easy to spot. They are the bad things we do. Harrelson's interpretation of *lassaw* is like sins of commission. We take God's name in vain when we do acts of evil, harm, and destruction. Such an interpretation would yield a rather simple set of ethics.

The broader interpretation, however, moves us to a much higher realm of ethical inquiry. To see the word as implying "lightly" or "with idleness" certainly addresses those "sins of commission", but now it brings in the "sins of omission." These are the things we fail to do; the purposes of God that we fail to carry out; the values of God that we fail to enact.

We have met, then, the stipulation of our freedom to harness the powers of God over nature. We can employ those powers freely, so long as we do not do so in vain. This certainly means that we should not utilize those powers for evil purposes, but that is so obvious it can almost go unsaid. Beyond that, however, is the real challenge of this commandment. We have been given the permission by God to possess God's powers over nature, *so long as we preserve God's purposes and God's values.* This moves us to our next great challenge, which will be the focus of the next chapter. Before we move on, however, let us stop to assess our journey thus far.

A Summary of Our Progress

We began this study with the recognition that our advancing abilities in biotechnology are presenting our faith perspective with new ethical challenges that seem to be outpacing our traditional ethics and theology. This has left us with difficult questions concerning the sacredness of physical matter, and of biological life. Many religious reactions to biotechnology have been quite negative, in which clear lines of demarcation have been drawn between nature that is sacred and that which is not. Cloning, stem cell research, animal-human hybrids, human-machine hybrids, etc…have all been seen as beyond the sacred line, and therefore off limits to human activity.

In our study, however, we have gone back to look at the notion of sacredness within the Judeo-Christian faith. We discovered that one of the chief differentials between the faith of the ancient Israelites and that of surrounding cultures is that this emerging faith broke from the ideas of polytheism and pantheism. In the faith of the ancient Israelites, there was one God, and that one God was separate and transcendent from nature. As such, no physical matter (that is, nothing in nature) was considered to be sacred in and of itself. In our faith tradition, there is no intrinsic sacredness in nature.

Our study went on to realize that when we employ this concept of nature, we must include ourselves in the discussion. Humanity is a part of nature. Our DNA is as much a part of nature as is the dirt of our gardens. None of it is intrinsically sacred.

We also learned, however, that sacredness can be a quality imposed on nature. Something becomes sacred when one sets it apart, and claims it for one's own use. When God sets a part of nature aside, and claims it for God's own purpose, then that part of nature is sacred unto God. As we considered the doctrine of general revelation, as well as the emerging theology of the world as sacrament, we arrived at the conclusion that in fact, God has set apart all of nature for God's purpose of self revelation and for the extension of God's grace. None of nature is sacred unto itself. All of nature is sacred to God.

This brought us to the analogy of traffic lights. If all of nature was intrinsically sacred, the light for biotechnology to encroach on nature's space would be red. We could find no justification to proceed. If nature was not sacred in any sense, the light would be green. Biotechnology would have every right to proceed into the deepest spaces of biology and life. Seeing nature as sacred to God, however, offers a yellow light. In this case, our encroachment into the sacred space of nature is not an inalienable right, but can only happen by God's gracious permission. In giving that permission, God can establish any stipulations that God chooses.

Our next task was to search in the Scriptures of our Judeo-Christian faith to see if God granted any such permission, and if so, if any stipulations accompanied that permission. We were drawn to the foundational text of the Decalogue, which truly stands as a basic pillar of our entire faith tradition. Before approaching the Decalogue itself, we observed the symbolic meaning behind the ancient concept of taking the name of a deity, and we observed in both the surrounding Egyptian culture as well as the Israelite culture, the mythical understanding that to take the name of a deity meant to possess the power of that deity. To take the name of the God of Israel, then, would offer humanity access to God's power over nature. We then turned to the Decalogue to witness a grand act of permission concerning the name of God.

Before considering the actual commandment, we made some important contextual observations about the entire Sinai covenant. We recognized that it was born out of a spirit of liberation, and of a new relationship between God and his people. The overwhelming spirit of the covenant is one of freedom, and the commandments themselves serve as basic stipulations to keep the exercise of that freedom within rightful bounds. With this understanding, we recognized that in the preamble of the Decalogue, God gave his name. The freedom to take hold of God's power over nature was on the table. In the third commandment, God stipulated that freedom. We were told that we could not take God's name in vain. This led us to make the important statements that are repeated here: We have been given the permission by God to possess God's powers over nature, *so long as we preserve God's purposes and God's values.*

As we proceed in this book, then, we will approach each chosen aspect of biotechnology with the basic and bold conviction that we are operating within the realm of freedom. It is not a freedom born out of our rights, but rather out of God's permission. We will never claim that something should not be done because it is tampering with sacred space. Instead, we will ask of each effort this simple question: Is this act preserving God's purposes and God's values?

In order to appropriately answer such a question, we must find a more concrete understanding of what "God's purposes and God's values" might mean. To the degree that it is possible, it would be helpful to focus this rather nebulous concept and to establish some basic markers that would serve as a guide, a map as it were, as we seek to navigate our stipulated freedom. Perhaps it would be too ambitious to suggest that we could draw a map. A map is quite exact, and one would have to claim a great deal of knowledge of a particular territory in order to draw one. This path of biotechnology is quite uncharted, and the exact will of God is always partly mystery. If aiming at a map is too high, perhaps a better metaphor would be a compass. A compass may not be as exact as a map, but it certainly becomes a helpful tool when one wishes to navigate a journey. A compass points into a general direction. It may not tell us what lies ahead, but it will tell us if we are pointed in the direction of a certain goal.

Once such a compass is developed, we could then travel this road, and see what guidance it offers. Aligned to the magnetic pull of God's purposes and values, this compass would point us in the general directions of ethical choices for the uncharted path of biotechnology.

What are the main polarities of this compass? This becomes the important question for the next chapter.

Chapter 2 Notes

[59] Exodus 3:13

[60] This theme is well articulated in the book The Gift of the Jews. *How a Tribe of Desert Nomads changed the Way Everyone Thinks and Feels*, by Thomas Cahill.

[61] Hornung, Erik. Conceptions of God in Ancient Egypt. *The One and the Many*, p.86

[62] Erman, Adolf. Life in Ancient Egypt. (Translated from the German by H.M. Tirard), p. 265

[63] Ibid, p. 265

[64] Hornung, Erik. Conceptions of God in Ancient Egypt. *The One and the Many*, p.88

[65] Genesis 32:29

[66] Judges 13:18

[67] Propp, William C. The Anchor Bible Commentary on Exodus 1-18, p. 224

[68] Ibid, p. 225

[69] Bennett, W. H. (Ed). The New Century Bible Commentary on Exodus, p. 58

[70] Driver, S. R. The Book of Exodus, p. 363

[71] Lunbhom, Jack R. *God's Use of the Idem per Idem to Terminate Debate*. In the Harvard Theological Review, vol. 71. 1978 p. 199

[72] Propp, William C. The Anchor Bible Commentary on Exodus 1-18, p. 224

[73] Brown, William P. (Ed) The Ten Commandments. *The Reciprocity of Faithfulness.*, p. 14

[74] Lochman, Jan Milic. Signposts to Freedom. *The Ten Commandments and Christian Ethics.*, p. 13

[75] Ibid, p. 17

[76] Von Rad, Gerhard. Old Testament Theology. Vol. 1. *The Theology of Israel's Historical Tradition*, p. 201

[77] Ibid

[78] Greeley, Andrew M. The Sinai Myth, p.51

[79] Von Rad, Gerhard. Old Testament Theology. Vol. 1. *The Theology of Israel's Historical Tradition*, p. 193

[80] Stamm, Johann Jakob. The Ten Commandments in Recent Research. (Translated from the German by Maurice Edward Andrew), p. 43

[81] Greeley, Andrew M. The Sinai Myth, p.48

[82] Ibid

[83] Lochman, Jan Milic. Signposts to Freedom. *The Ten Commandments and Christian Ethics.*, p. 28

[84] Harrelson, Walter. The Ten Commandments and Human Rights, p.20

[85] Von Rad, Gerhard. Old Testament Theology. Vol. 1. *The Theology of Israel's Historical Tradition*, p. 194

[86] Matthew 22:37-40

[87] Kohlenberger, John R. (Ed). The NIV Interlinear Hebrew-English Old Testament. Volume 1/Genesis-Deuteronomy, p. 200

[88] Keck, Leander, et al (Editorial Board) The New Interpreter's Bible. Volume 1, p.841

[89] Durham, John I. Word Biblical Commentary Volume 3. Exodus, p. 284

[90] C-Span.org

---

[91] Von Rad, Gerhard. Old Testament Theology. Vol. 1. *The Theology of Israel's Historical Tradition*, p. 194

[92] Lochman, Jan Milic. Signposts to Freedom. *The Ten Commandments and Christian Ethics.*, p. 20

[93] A central theme in Peter's book Playing God? *Genetic Determinism and Human Freedom*

[94] Exodus 20:7

[95] Moehlman, Conrad Henry. The Story of the Ten Commandments. *A Study of the Hebrew Decalogue in its Ancient and Modern Application.*, p.117

[96] Harrelson, Walter. The Ten Commandments and Human Rights, p. 72

[97] Herbert Huffman. The Ten Commandments. *The Reciprocity of Faithfulness*, (Brown, William P, Editor) p.207

[98] Harrelson, Walter. The Ten Commandments and Human Rights, p. 73

Creating the Compass

Introduction

One of the social organizations found in many communities is the Rotary Club. The Rotary Club is a social organization that unites community leaders for meals, socializing, and to make significant contributions to the betterment of the community and the world. In the typical meeting of a Rotary Club, there is a regular routine. When the bell rings, members will stand to say the Pledge of Allegiance, and then pause for an invocation. Following the invocation, members will recite together a memorized script that, in Rotary, is known as the *Four Way Test*[99]:

> Is it the truth?
> Is it fair to all concerned?
> Will it build good will and friendship?
> Will it be beneficial to all concerned?

If one is unfamiliar with this practice of Rotarians, one might ask a question about the four way test. What is the *"it"* that the test addresses? Actually, that is just the point. "It" is undefined. It is any idea, any plan, any proposal one would make. It is every business decision, every action. In assessing the value of a proposed action, Rotarians are encouraged to submit the proposed action to this test.

The test becomes a kind of ethical compass. If the proposed action passes the test of these statements, then the compass shows that the intended action is in the right direction, and is thus worthy of Rotarian support. If a proposed action fails to pass these statements, the compass steers away from that action, and it is deemed unacceptable for the faithful Rotarian.

Such an ethical compass can also be seen in the New Testament book of Philippians. As the Epistle's author was ending the letter, he knew that he could not address every issue of ethical living for the Christians in Philippi. In place of an exhaustive list of right ethical responses, the author gives his readers something else. It is a compass, by which they could navigate the pathway of any issue and determine its ethical measurement. He writes:

> *Finally, brothers and sisters, whatever things are true, whatever things are noble, whatever things are just, whatever things are pure, whatever things are lovely, whatever things are of good report, if there is any virtue and if there is anything worthy of praise, think on these things.*[100]

The real meaning of these words isn't totally reflected in most of our English translations. The author isn't simply encouraging positive thinking, as it may seem. The real intent of the Greek word λογιζεσθε isn't just thought, but rather, the thought process that determines action. It is an ethical compass. Proposed actions should be measured against these concepts and if those proposed actions are able to pass this test, then the actions can be realized.

These two examples illustrate the important task of this chapter. We have already seen evidence that God has initiated a tremendous freedom to humanity, and has invited us to participate in the powers that were once considered God's sacred space. But the freedom is not total, and not unrestricted. It is a stipulated freedom. We shall not harness God's power in vain. We must ensure that God's values and God's purposes are never compromised as we take hold of God's power. It is the stipulation of our freedom that necessitates a compass. In this chapter, we will make a humble attempt at developing such a compass. Our first order of business in developing this compass is to find a more concrete understanding of what we have, to this point, called God's "purposes and values." We will attempt to focus the broad scope of this statement into two overarching themes of God's activity as represented in the Scriptures. The articulation of these two themes does not mean that God has only two values, to be sure. God has many values, just as a person or an organization may have many values. Yet, in focusing the purpose of a life or an organization upon some limited, overarching themes, one can better define and understand the total set of behaviors and choices of that individual or organization. In organizational theory, we call these "core values." Without inflicting the dynamic personality of God with an assault of reductionism, is it possible to glean from the Judeo-Christian Scriptures a sense of God's core values? If so, then we would be well on the way to understanding the stipulations that God has placed on the human manipulation of nature. With that knowledge, we could proceed to develop the compass and apply it to evaluate several broad areas of emerging biotechnology.

God's Core Values

Attempting to assert and defend a few "core values" of God as expressed in the Judeo-Christian Scriptures is a daunting task. One could begin this search from a variety of angles and directions. Perhaps here it would be good to choose one such angle, and then look at a few representative others to see if the various sources and interests agree.

One assessable source of evaluating the biblical core values of God is offered by the growing number of biblical scholars who believe that those core values are being lost and distorted by modern American Christianity, specifically by the politically powerful "religious right." In an attempt to recover the "real God", or the "real Jesus" of the Bible, these scholars offer an investigation into an understanding of the guiding purposes and values of God as understood in the Bible. Let us consider two of the leading books on this subject: Jesus Against Christianity. _Reclaiming the Missing Jesus_, by Jack Nelson-Pallmeyer[101]; and The Politics of Jesus. _Rediscovering the True Revolutionary Nature of Jesus' Teachings and How They Have Been Corrupted_, by Obery M. Hendricks Jr.[102]

Nelson-Pallmeyer begins his book by grounding an understanding of Jesus' ministry upon the foundation of the Hebrew Scriptures, ancient Judaism, and the ancient Jewish nation. He believes that, indeed, one core value permeates through all of that. It is the theme of _justice_. The core activity of this God of the Hebrews, according to

Nelson-Pallmeyer, is established in the Exodus event.[103] The Exodus event and the Sinai Covenant (as we observed in the last chapter) illustrate the fundamental work of God as the liberator of the oppressed. From that pivotal moment, this central theme of justice becomes the beating drum of a faithful human community in proper relationship to God.

Biblical justice, according to Nelson-Pallmeyer, is best understood in terms of *equality*. Justice is realized when the resources of the community are properly and evenly distributed among the population of the community. This drive toward equality, and of the balancing of resources, is at the heart of the Sabbath, particularly expressed in the Sabbath Year, or the Year of Jubilee. Nelson-Pallmeyer writes:

> How were people to respond to a God of Justice who delivered them from oppression? How were they to structure their life together once they gained control of their own land? The answer seemed clear. A just God required the people of God to take on the vocation of justice. Their social order was expected to reflect, promote, establish, and maintain justice and equality. Justice was possible when land ownership was distributed evenly, interest forbidden, and debts periodically canceled. At set intervals, the cycles of poverty were to be interrupted or broken. Laws that define the people's proper relationship to land, debt, interest, and slavery were at the heart of the covenant.[104]

In summing up his sense of the "core value" of God as expressed in the Hebrew Scriptures, Nelson-Pallmeyer states that "God's justice is central to the character of God, and the people of God are expected to embody justice and equality in their social relationships."[105]

As Obery Hendricks Jr. begins his book of rediscovering the true ministry of Jesus and setting it free from modern day distortions, he too begins by grounding the ministry of Jesus in the ancient Hebrew world. Like Nelson-Pallmeyer, Hendricks believes that the character of God, and of God's intended relationship with Israel, is formed and sealed in the Exodus event.[106] From there, Hendricks traces the central emphasis of each period of ancient Hebrew history, through the time of the Judges, the kingly and Messianic rule, the eighth century prophets, the exile, the Maccabean history, the Herodian rule, and right up to Jesus own day. Like Nelson-Pallmeyer, it is Hendricks' conviction that the overarching value representative of God's character is captured in the theme of *justice*. Hendricks points out that there are two main words in the Hebrew Scriptures for the word "justice." The first word is משפט (*Mishpat*). This word connotes the sense of equality. Hendricks writes that this word indicates:

> ...the establishment or restoration of fair, equitable, and harmonious relationships in society. The major implication of its meaning is that any member of the community has the same rights as any other, that everyone has the same inalienable right to abundance and wholeness and freedom from oppression. *Mishpat* also means "judgment" in the sense of balancing or working to resolve all conflicts-social, economic, and political-with the equal rights of all in mind.[107]

The second main word that is translated into English as "justice" is the word צדקה (*Tzedakah*). In many versions of the Bible, this word is translated as "righteousness" rather than "justice". According to Hendricks, the modern understanding of this word has, itself, become distorted, as it is often understood as a matter of an individual relationship with God. He states that the real meaning of the word indicates a fulfillment of relationship with God *and* with others. Hendricks writes:

> The basis of biblical justice (*tzedakah*) is fulfillment of our responsibilities to and relationships with others as the ultimate fulfillment of our responsibility to God.[108]

These two writers, representative of a larger body of writers who are seeking to set the record straight concerning the core values of the Bible amid the misguided and distorted images of modern American Christianity, have converged on one central core value that is representative of God, and of the person and work of Jesus Christ. It is the value of *justice*. This justice is realized when the resources of the world are evenly distributed among the populations of the world. It is realized when the abundance of the world becomes everyone's equal right. Moreover, we achieve righteousness with God when we have fulfilled our responsibilities of sharing the abundance of resources with others. This core value of justice, understood in these terms, has some challenging implications for the ethical exercise of biotechnology. Far from restricting efforts of biotechnology, this core value may actually encourage bigger and bolder steps. Understood as a stipulation, however, requires that these big and bold steps never leave a portion of the world's population behind. With this, we begin to get some clarity about the compass that we are seeking to develop. The polarities that lead us in the direction of God's purposes and values are beginning to come in focus. We will get started on the development of this compass in due course. First, there is a question to ask, and one more core value to explore.

The question to be asked is whether the authors we have considered have it right. The purpose of this question is not to doubt the scholarship of these two writers. Both men have good credentials, and we can be confident that they have developed their arguments on good and solid scholarship. It is also apparent, however, that both are guided by a particular agenda. They want to expose what they consider to be the misguided faith of the religious right. As stated earlier, this is but one angle to address a major question. Once we have this view, it is important to test it by comparing the conclusions with a few other scholars who do not have this same agenda. Do other scholars, who approach biblical theology from a more neutral position, come to the same conclusion? In their perspectives, is justice a core value of God?

As I ask this question, my memory floods back to my seminary years, in which I had the wonderful privilege of learning Biblical theology, particularly that of the Old Testament, under the professorship of Dr. Walter Brueggemann. Walter Brueggemann is widely regarded as one of the world's leading authorities on Old Testament theology. His classroom lectures left a lasting impression on his students, and often transformed our thinking. In his lectures, Brueggemann made it immensely clear: God is for the poor; God is about hope for the disenfranchised; God is about *justice.*

Brueggemann certainly understands the complexity and diversity of biblical themes. Even though it may risk the accusation of reductionism, Brueggemann asserts, and aptly defends, his view that justice for the poor and disenfranchised is the predominant value of the God we meet in the Hebrew Scriptures. In his Old Testament Theology, Brueggemann writes:

My thesis is this: In the face of the rich pluralism and passionate interestedness of the biblical text in its various local voices, the text everywhere is concerned with the costly reality of human hurt and the promised alternation of evangelical hope.[109]

Bringing "evangelical hope" to mitigate "human hurt" is the essence of God's desired justice for the world, and God's people find their righteousness in relationship to God when they extend God's hope to the hurting of the world. In order to do this, the abundance of resources available must be properly distributed to all. This, for Brueggemann, is the meaning of *justice*. Once again, it carries the sense of equality. In his book for pastors entitled <u>To Act Justly, Love Tenderly, Walk Humbly: *An Agenda for Ministers*</u>, Brueggemann writes:

> There are, of course, various and conflicting understandings of justice. Let me offer this as a way the Bible thinks about justice: Justice is to sort out what belongs to whom, and to return it to them. Such an understanding implies that there is a right distribution of goods and access to the sources of life.[110]

From another reliable source from a much different perspective, then, we find confirmation that in searching for a "core value" to express the character and purposes of God, the concept of justice comes in focus.

Let us consider one more source to solidify the point. In his classic book <u>The Hebrew Bible: *A Socio-Literary Introduction*</u>, Norman Gottwald spends some time focusing on the prophets of the Old Testament. When one reads the prophets, it becomes clear that social justice for the poor is a major theme. The prophet Amos is presented as a particular and poignant example of this. Amos was a rural prophet, and most likely had no formal training in either the theology or the politics of Judaism. From what imperative, Gottwald asks, does Amos conclude that injustice is the central sin of Israel? He writes:

> If we pay heed to the crushing of the poor as the central sin Amos condemns and to the way he pictures divine activity

through rhetorical questions and figures of speech from the natural and social world of rural Palestine, we can make an informed estimate of what weighted most in his thinking. He knew firsthand about the murderous oppression of the poor; not only did he detest that oppression, but he knew that it was diametrically opposed to Yahweh's wishes. How did he know this? He knew it from the traditions in narrative and law shaped by the old tribal life of Israel and presently enshrined in the practice of mutual help and discharge of justice in local courts. The substance of the socioethical laws of Israel was known to him, even if he had never seen written laws.[111]

In other words, Amos knew that justice was a core value of God even though he had no formal training in theology because this knowledge was clear and unquestioned in the cultural life of Israel, even in the most rural areas. What was clear to ancient Israelites is becoming clear to us. Justice, the balancing of resources with the entire population, is a central and core value of God. If this has been lost in popular America Christianity, then it is important that writers such as the two who began our search for God's core values get heard. A Christianity based on wealth and political strength is certainly out of step with the God we meet in the Bible.

An Application to Our Stipulated Freedom

The God we meet in the Bible, or at least in the Hebrew Bible, has the purpose and the value of justice. When we use our talents and abilities to promote the cause of justice in the world, then we are acting in accordance with God's purposes and values. When we purposefully participate in acts of injustice, we are clearly acting in opposition to God's purposes and values. In doing such, the Bible would call us "unrighteous". Let us remember, however, what Obery Hendricks said about the Hebrew word (*sadiqah*):

> The basis of biblical justice (*sadiqah*) is fulfillment of our responsibilities to and relationships with others as the ultimate fulfillment of our responsibility to God.[112]

Our righteousness is secured not merely in the restraint of injustice, but also in our failure to fulfill God's purpose of justice. In other

76

words, it is not just a matter of *commission*. It is also a matter of *omission*.

In this book, we have been exploring a biblical ethic for the practice of biotechnology. We have recognized that nature (all matter) is not intrinsically sacred, but has been made sacred by God. We have also seen that God invited us to exercise power over nature, as long as in doing so, we preserve God's purposes and values. This is expressed in the commandment "Thou shall not take my name *in vain*". Understood in the light of the core value of justice, this stipulation begins to take shape. God invites us to seize power over nature, as long as in doing so, we balance the benefits among the entire population. A manipulation of nature that results in a benefit for all (or at least is moving toward that goal) would be allowed by this stipulated freedom. More than that, such manipulation would even be encouraged. One can come to this conviction when the Decalogue is seen in its proper light. God begins with freedom, and an encouragement to seize that freedom. Only after that freedom is asserted, God moves to establish the stipulations. This gives a bold imperative to biotechnology. If, indeed, our righteousness is found in fulfilling the quality of life for the world's disenfranchised, a biotechnological creation that is far removed from the original state of nature yet offers a positive benefit and improvement that is, at least theoretically, available for all would pass the test of our stipulated freedom.

We must ask, however, what aspects would not pass this test. What efforts of biotechnology might be taking God's name in vain? Obviously, any manipulation of nature that promotes injustice would fall into this category. Technology that has, as its main function, the purpose of bringing harm and increasing levels of hurt would fail the test. Manipulating the smallpox virus in order to create a vaccine to protect the world's population from this disease is a wonderful use of God's stipulated freedom, and a positive step toward righteousness. Manipulating the same virus, however, to create a weapon for germ warfare is clearly not. Such an act breaks the commandment. It shatters the covenant. It trespasses on Holy Ground.

Manipulations of nature that have a clear purpose of harm are the easiest ones to spot among those that take God's name in vain. There are others, and that is where the stipulated freedom becomes more difficult to navigate. What about biotechnological activity that does not set out to bring harm, but results in an unintended element of hurt to a segment of the world's population? On the very day that these words are being written, the Intergovernmental Panel on Climate Change has released its 2007 report. The report strengthens the scientific claim that human activity, primarily the use of fossil fuels, is doing significant harm to the earth's climate. The purpose of the technology that converts natural elements into fuel did not purposely set out to do harm to the earth or its inhabitants. Yet, an unintended element of hurt has developed. How would one evaluate this in terms of our stipulated freedom?

The compass we are seeking will have to take this question into consideration. Likewise, the compass will have to offer guidance on other aspects of biotechnology that fail to meet the imperative of justice. As we will see in the upcoming chapters, many emerging possibilities of biotechnology seem to fit more into the realm of consumer commodities rather than generalized human needs. As such, these aspects of biotechnology will be considered as a luxury rather than a therapy. They will not be covered by most insurance programs. There will be no drive to offer them to those who are unable to afford them. These aspects of biotechnology will dig a deeper divide between the world's "have's and have not's". Today, this dynamic is experienced in terms of cars and computers, leading to what some have labeled the "technological divide", or the "digital divide." In tomorrow's world, this divide may be measured by such basic realities of nature such as intelligence, functional bodies, and life spans with differences measured not in decades, but perhaps centuries. If such possibilities become actualized, how would they measure up to God's stipulated freedom? Where is justice to be found?

We have raised questions about technology that purposely promotes harm, technology that has unintended consequences of harm, and technology that provides an unbalanced benefit to the world's rich. Before moving forward, there is one other question that must be raised. What about biotechnological activity that, for all purposes, is neutral in terms of justice? Xenografting technologies, for example, can have many therapeutic benefits. This technology, which will be discussed more fully in the next chapter, involves grafting tissues from one species to another in order to create some level of life that is a hybrid between the two. Scientists working in these fields believe that these hybrid creations could lead toward the development of tissues and organs that can be utilized in the treatment of many diseases. The same technology, however, could be utilized for a myriad of experiments that have no purpose other than to prove that something can be done, and push beyond the limit of previously accepted norms. This became an interesting debate back in the late 1990's when Dr. Stuart A. Newman, professor of cell biology and anatomy at New York Medical College, applied for a patent on a "humanzee", a hypothetical creature that would be half-human and half-chimpanzee. His patent was denied, amid much controversy. In attempting to explain why he wanted to pursue this patent, Dr. Newman explained that he wanted to protect this technology from those who may want to exercise it simply to be the first to do so. In a printed interview with the *New York Times* published May 31, 1998, Newman said:

> I think that the issue we're raising by doing this is that there must be limits to this kind of work. It can't be done in a completely unregulated fashion. It can't be driven only by scientific curiosity and commercial viability. There have to be some other values that are brought to bear on this kind of enterprise.[113]

Given the conclusions of our study thus far, what would we say about this strange case? On the one hand, we would have to conclude that the actual manipulation of nature to create a human-chimpanzee hybrid is *not*, in and of itself, contrary to God's stipulated freedom. Yet, we would also agree with Dr. Newman that such efforts can not simply be driven by "scientific curiosity". Manipulations of nature that are neutral to God's core value of justice are utilizing God's power without God's purposeful intent. These efforts would be taking God's name in vain.

In presenting these parameters around the exercise of biotechnology in relationship to God's core value of justice, we have identified some key components in the development of our compass. Before we move forward in that task, there is one other core value that we must consider. Christianity, the religious context from which this book is written, is based on the entire Bible, not just the Hebrew Scriptures. It is the Christian belief that the God of the Old Testament becomes more fully known in the pages of the New Testament, and in the person of Jesus Christ. As we turn to the New Testament, do we see continuity with the major themes that we have observed primarily in the Hebrew Scriptures? Is the theme of a stipulated freedom evident? Is the core value of justice present? In the following section of this chapter, we will consider these questions. As we do, we will discover another core value of God. It is closely related to the first, yet develops a new level of looking at God's stipulated freedom. Let us now turn to the New Testament, and consider the core value of *love*.

God's Stipulated Freedom in Light of the New Testament and the Person of Jesus Christ

As we now turn our attention to the New Testament, we will begin by testing the main hypothesis of this book against the writings of the New Testament, and the personal ministry of Jesus Christ. In the Sinai covenant, God established a context of freedom, and then presented the parameters of that freedom with stipulations. The stipulations, as we have noticed, were not about the elements of nature, but rather, were about the human community. How do these statements compare to the theology of the New Testament?

First, there is the ministry of Jesus Christ. As we noted in the last chapter, Jesus approached the law of the Old Testament in a rather unique way for the time. He encouraged his followers, for instance, to go against the generalized prohibitions of the Sabbath. In one such episode, recorded in Mark 2, we read:

> One Sabbath Jesus was going through the grainfields, and as his disciples walked along, they began to pick some heads of grain. The Pharisees said to him, "Look, why are they doing what is unlawful on the Sabbath?" He answered, "Have you never read what David did when he and his companions were hungry and in need? In the days of Abiathar the high priest, he entered the house of God and ate the consecrated bread, which is lawful only for priests to eat. And he also gave some to his companions." Then he said to them, "The Sabbath was made for man (sic), not man for the Sabbath."[114]

In this episode, several points of congruency can be observed between Jesus' teachings and God's stipulated freedom found in the Decalogue. On the surface, it may seem that Jesus is subverting the commandments, as was charged of him by the Pharisees. In reality, we are reminded of the writings of Gerhard Von Rad and others, who have argued that the original purpose of the Decalogue was to establish the broad limits of human freedom so that the freedom could be exercised in a responsible way. In this episode from Mark, Jesus is seeking to restore this aspect of freedom to one of the commandments, which calls for the honoring of the Sabbath. It is also interesting to note that Jesus chose to include in his argument the Old Testament example of David eating the "consecrated bread" that is reserved for the priests. For Jesus, bread, as an element of nature, had no intrinsic value of sacredness. It is made sacred by God, and God, therefore, has the right to permit human encroachment.

This episode is but one of many stories contained in the Gospels in which Jesus purposely acts with freedom in regards to the Sabbath laws. In many cases, Jesus invites his disciples to follow suit. This offered to the disciples of Jesus a newfound sense of religious freedom. This freedom, however, was not a *right* of the disciples. It was, rather, a *permission*. In his <u>Theology of the New Testament</u>, the German scholar Leonhard Goppelt writes:

> Those who followed him in discipleship also participated in this freedom of Jesus. This can be seen in the narrative about plucking grain on the Sabbath. Here it was not the conduct of the disciples as such that justified but the fact that Jesus permitted it.[115]

When we move from the actual teachings and ministry of Jesus to the foundational texts of the Christian faith, we come to the writings of the Apostle Paul. It is in Paul's writings that the freedom of Christian theology becomes more fully developed. One can gain a good understanding of Paul's view of Christian freedom in the book of I Corinthians. The Corinthians were facing a moral dilemma. In converting to the Christian faith, they had abandoned the worship of the many idols that dotted the landscape of ancient Corinth. They knew that the worship of idols was morally forbidden. They also believed that anything derived from the worship of idols would be morally tainted, and therefore untouchable. One of the things that were derived from the worship of idols was meat. Animals were regularly sacrificed at these idols, and their meat entered the food chain of Corinth. This meat, in the view of the Corinthian Christians, was morally tainted. It should be avoided. Yet, this morally tainted meat was a common benefit throughout the city. It ended up in the meat markets, and at social gatherings. A Corinthian Christian would often not be able to know if the meat being offered comes from the idols or not. In some communication to the Apostle Paul, the Corinthian church asked for guidance. Through a large section of this New Testament Epistle, Paul addresses this question. As he does, Paul asserts two main points. First, Christians have a *freedom*. He speaks of "this liberty of yours" in I Corinthians 8:9; and in a first person reflection, spends most of chapter 9 defending this freedom. Because of this freedom, Paul states, the Corinthian Christians can, in fact, eat the meat without worrying about the moral implications. The meat, according to Paul, has no moral value. It is not sacred, nor is it morally tainted. This is an important point of Paul's theology to which we will return in a later chapter, as we discuss the issue of stem cell research. For now, it is enough for us to simply observe that Paul asserts a Christian freedom over the nature of the meat.

This freedom, however, is not absolute. There is something that qualifies it. This is Paul's second main point. It can be observed in this passage:

> "Everything is permissible"—but not everything is beneficial. "Everything is permissible"—but not everything is constructive. Nobody should seek his own good, but the good of others. Eat anything sold in the meat market without raising questions of conscience, for, "The earth is the Lord's, and everything in it. "If some unbeliever invites you to a meal and you want to go, eat whatever is put before you without raising questions of conscience. But if anyone says to you, "This has been offered in sacrifice," then do not eat it, both for the sake of the man who told you and for conscience' sake -the other man's conscience, I mean, not yours. For why should my freedom be judged by another's conscience? If I take part in the meal with thankfulness, why am I denounced because of something I thank God for? So whether you eat or drink or whatever you do, do it all for the glory of God. Do not cause anyone to stumble, whether Jews, Greeks or the church of God—even as I try to please everybody in every way. For I am not seeking my own good but the good of many, so that they may be saved.[116]

For Paul, the Christian life is one of liberty, curtailed by a predominant value. It is a stipulated freedom. The stipulation of this freedom is not about nature (i.e. meat sacrificed to idols), but about the human community. The Corinthians were encouraged to exercise freedom over the material world so long as they did not violate a core value concerning others. This core value is not so much that of justice, but of the one we are soon to discuss: the core value of love.

In both the ministry of Jesus and in the writings of the Apostle Paul, we do see congruency with the concept of a stipulated freedom. In both cases, we can conclude that the stipulations do not find their locus in the material world. They are grounded, rather, in the core values of God in God's care for the human community. In our consideration of the Hebrew Scriptures, we observed the preeminence of justice as one such core value. Does justice find equal import in the New Testament?

Justice as a Core Value in the New Testament

One of the defining moments of Jesus' identity and purpose can be seen in Luke 4. Returning from his trials in the wilderness and about to begin his public ministry, Jesus returns to his childhood town of Nazareth. While in the synagogue, Jesus stood up to read the scripture of the day, a passage from Isaiah 61:

> The Spirit of the Lord is on me, because he has anointed me to preach good news to the poor. He has sent me to proclaim freedom for the prisoners and recovery of sight for the blind, to release the oppressed, to proclaim the year of the Lord's favor.[117]

After reading this text, Jesus told the congregation that Isaiah's passage was being fulfilled before their eyes. From Jesus' own lips, we hear the focus of his ministry. He came to do the work of justice. When one stops to consider the *actions* of Jesus as he healed the lame, ate with outcasts, associated with lepers; and when one stops to consider the *teachings* of Jesus as he preached about the widow's mite, the Good Samaritan, the rich lawyer who was told to sell everything and give to the poor; it becomes clear that much of his ministry was about social justice. When we began our discussion on the core value of justice in this chapter, we turned to two books that initially pointed us in this direction. These books, as you recall, were not about the Hebrew Scriptures. They were about the ministry of Jesus Christ. The consideration of the Hebrew Scriptures was offered to show the foundation upon which Jesus' ministry was built. Both The Politics of Jesus, by Obery M. Hendricks Jr. and Jesus Against Christianity, by Jack Nelson –Pallmeyer, make a solid case for understanding Jesus' ministry as solidly built on the core value of justice.

Throughout the remainder of the New Testament Scriptures, we see the theme of justice sprinkled through the pages. The New Testament book of James, as one example, shouts for justice:

> Now listen, you rich people, weep and wail because of the misery that is coming upon you. Your wealth has rotted, and moths have eaten your clothes. Your gold and silver are corroded. Their corrosion will testify against you and eat your flesh like fire. You have hoarded wealth in the last days. Look! The wages you failed to pay the workmen who mowed your fields are crying out against you. The cries of the harvesters have reached the ears of the Lord Almighty. You have lived on earth in luxury and self-indulgence. You have fattened yourselves in the day of slaughter. You have condemned and murdered innocent men, who were not opposing you.[118]

Justice is a core value, to be sure. It was a core value of the Hebrew Scriptures, and of the nature of God as revealed in them. It was a core value of Jesus, represented by both his words and his actions. It was a core value of the early church, represented by other New Testament writings.

With an observance of the New Testament teachings on the themes of freedom and justice, we can feel confident that the Christian Scriptures confer with the conclusions of this book thus far. In addition to confirming what we have already observed, however, the New Testament adds a heavily weighted value that must be included before we consider the compass. It is time to invite this value into our journey. Let us consider the second core value to be added to our compass: the core value of *love*.

A Second Core Value: Love

It isn't a difficult task to demonstrate that love is a core value of the New Testament. The point is made by Jesus himself when he is asked, in a sense, "what is God's core value?" This exchange can be witnessed in this passage from Matthew:

> Hearing that Jesus had silenced the Sadducees, the Pharisees got together. One of them, an expert in the law, tested him with this question: "Teacher, which is the greatest commandment in the Law?" Jesus replied: "Love the Lord your God with all your heart and with all your soul and with all your mind. This is the first and greatest commandment. And the second is like it: 'Love your neighbor as yourself.' All the Law and the Prophets hang on these two commandments."[119]

This two-fold love commandment, as it has been often called, has been seen by many as clear evidence that love is a primary, if not *the* primary, ethic of Jesus and his early followers. In Jesus and the Ethics of the Kingdom, co-authors Bruce Chilton and J.I.H. McDonald express this view aptly:

> The focus of the new obedience is found in the twin commandment to love. Its centrality is illustrated by the introductory question in the various redactions of the tradition. It is not adequate exegesis to dismiss such features as merely redactional: the redactions give important guides to interpretation. The debate is about supreme obligation as expressed in the law: which is tantamount to discussing the overriding priority for human life and destiny. There can be little doubt about Jesus' acceptance of the twin commandment as expressing the primary moral requirement of the law.[120]

Although Jesus is depicted as presenting the primacy of the love commandment in the synoptic Gospels, it is the Gospel of John that presents the value of love in a clearly focused and brilliant light. Of this Gospel, noted New Testament Scholar Rudolph Schnackenburg said that "St. John is not only the royal guardian of Christ's inheritance preserving his spirit but also a disciple of the Lord illumined by the Holy Spirit, giving added profundity to the commandment of love and raising it to be the ruling principle of Christian morality throughout all ages."[121]

As we move to the writings of the Apostle Paul, the ethic of love is in no sense diminished. One of the most recognized passages in the entire Bible, thanks to its prominence in wedding ceremonies, is Paul's poetic words of I Corinthians 13. At the end of that chapter, Paul shows us the core value of the Christian faith: "Now faith, hope, love; abide these three. And the greatest of these is love."[122] In his book Christian Morality. *Biblical Foundations*, author Raymond F. Collins sums up Paul's focus on love this way:

> It is certain that Paul's letters witness to the preeminent place accorded to love by early Christian preachers. Indeed, more than sixty percent of the New Testament occurrences of the word 'love' are to be found within the Pauline corpus. This linguistic evidence confirms one's initial impression that Paul must have been one who preached about love, even if it was the evangelist John who has provided us with the essential elements for our theological reflection on love.[123]

Collins demonstrates evidence that the "faith-hope-love" triad that Paul employs within his writings is most likely not of his own origin, but rather a saying from philosophy or theology that pre-dated Paul.[124] Paul's assertion of love as the greatest of the three adds the Christian perspective.

It seems quite clear that love is, indeed, the predominant core value of Jesus and of the New Testament Christian community. The weighted emphasis on the theme of love found in the New Testament leads us to add it to our understanding of God's "purposes and values" as we seek to develop a compass to guide us in the journey through God's stipulated freedom. Now, the acknowledgment of love as a core value must be supplemented with an understanding of what kind of love this core value is.

Agape Love Defined. In the Koine Greek, the language of the New Testament, there were four main words utilized for the concept of love. The deeper delineation of "love" into these words enables the reader to have a more specified understanding of the nature of the love being described. The Greek word ἔρως *(érōs)* connotes sensual, or romantic, love. The word φιλία *(philía)* depicted a friendship love, or the "dispassionate virtuous love" conceptualized by Aristotle.[125] A third word for love, στοργή *(storgē)* represented familial love, such as that existing between parents and children.

Although these words have some limited usage in the New Testament Scriptures, it is the fourth word that represents the love that is seen as God's core value, and the predominant ethic of the Christian community. It is the love known as ἀγάπη *(agápē)*.

Agape is a different kind of love than the others. Its primary difference is that it is not based on affection or emotion. Agape love is more measured by actions than feelings. Agape love is the one type of love that is purely lacking in self-interest. In fact, it is represented in the opposite: self-sacrifice. The online Encyclopedia Wykpedia defines Agape love this way:

> In biblical literature, its meaning and usage is illustrated by self-sacrificing, giving love to all--both friend and enemy. The word seems to contain more of a mental or intellectual element than the other Greek words for love. It is a rational love that is not based on total self-interest. By this a Christian is required to love (agape) someone who is not necessarily lovely or loveable. The Christian by God's grace and mercy is required to "love" someone that he may not necessarily like or love in the sense of having warm fuzzy emotional feelings toward. It is a love that acts in the best interest of the other person.[126]

Agape love, then, is a category of love that may seem a bit foreign to us. It is not about emotion. It is not about special feelings, or special people. Agape love is an ability to act selflessly in the best interests of others, even if those others are not known by us, or perhaps liked by us. As CS Lewis wrote in <u>The Four Loves</u>, "divine gift love (i.e. Agape love) in the man (sic) enables him to love what is not naturally loveable."[127]

Agape love becomes stronger, in fact, when it is offered to those who are the most removed from our other categories of love. The strength of Agape love is not measured in intensity of feeling, but rather in the depth of self-sacrifice and the extent of reach to the stranger, or even the enemy. When we are able to do that, we are participating in God's core value of love. As CS Lewis said, "Let us make here no mistake. Our gift-loves are really God like. And among our gift-loves those are most God like which are most boundless and unwearied in their giving"[128]

Added to justice, then, we now have a second core value of God. It is Agape love. Together, love and justice become the guiding polarities of the compass that will lead us in the evaluation of several biotech subjects. We have already considered the kinds of questions that the core value of justice would raise. What additional questions would the core value of love present?

An Application to Our Stipulated Freedom

God's stipulated freedom invites us to take hold of creative powers over nature as long as we preserve God's core value of Agape love. Does this raise any additional issues for us that were not already raised when we considered the core value of justice? In some ways, the concepts are identical. In both cases, concern is expressed for the "have not's" of the world. In both cases, equality of resources is desired. When one considers the demands of Agape love along with the content of social justice, one could easily conclude that you could not have love without justice. The love which requires a concern for the welfare of others, and that is measured in concrete action rather than internal emotion, would necessarily take action for justice.

This being the case, do we need to differentiate between the two? Why not accept the priority of justice and be done with it? The problem with burying love under the larger umbrella of justice is that it may lead one to hold the false belief that you can achieve justice *without* love. In a finite world of resources, there can be no true justice without Agape love. This fact raises new challenges for the exercise of biotechnology within the framework of God's stipulated freedom.

If the world were a place of infinite resources, then we could achieve justice by a one sided approach. We would simply work to raise the standard of life for others until they reach a level of equality with us. In this case, you could have justice without love. The balancing of the world's resources with the world's population could be achieved without any need to sacrifice. Our world, however, is not a place of limitless resources. There are limits at many different levels. There are limited dollars to be expended on biotechnological research and development. There are limited numbers of facilities to do this R&D, accompanied by a limited number of human resources to accomplish the work. At an even more basic level, there is a limit to the actual material resources available. There is only so much coal. There is only so much vaccine. There is only so much corn.

Corn is a good example to consider here. In his 2007 State of the Union Address, President George Bush emphasized that in order to protect our way of life, our nation will have to reduce its dependency on foreign oil. In order to accomplish this, he encouraged the future use of ethanol, a fuel derived by plant material, specifically corn. Bush said:

> Extending hope and opportunity depends on a stable supply of energy that keeps America's economy running and America's environment clean. For too long, our nation has been dependent on foreign oil. And this dependence leaves us more vulnerable to hostile regimes, and to terrorists who could cause huge disruptions of oil shipments ... raise the price of oil ... and do great harm to our economy. It is in our vital interest to diversify America's energy supply, and the way forward is through technology. We must continue changing the way America generates electric power by even greater use of clean coal technology ... solar and wind energy ... and clean, safe nuclear power. We need to press on with battery research for plug-in and hybrid vehicles, and expand the use of clean diesel vehicles and biodiesel fuel. We must continue investing in new methods of producing ethanol — using everything from wood chips, to grasses to agricultural wastes.[129]

Investing in new methods of producing ethanol from "wood chips, to grasses to agricultural wastes" is a clever way to avoid the controversial aspect of ethanol production. Lester Brown, of the Earth Policy Institute, has highlighted key concerns about our move toward ethanol. First, Brown articulates the realities. At the end of 2005, Brown wrote:

> Cars, not people, will claim most of the increase in world grain consumption this year. The U.S. Department of Agriculture projects that world grain use will grow by 20

89

million tons in 2006. Of this, 14 million tons will be used to produce fuel for cars in the United States, leaving only 6 million tons to satisfy the world's growing food needs.[130]

According to Brown, the reduced amount of grains available for food supplies will present two challenges. First, the scarcity itself will result in less food sources available for the world's population. Second, the increasing demand for grains in the highly profitable fuel industry will result in higher costs for those grains. Brown states:

> As the price of oil climbs, it becomes increasingly profitable to convert farm commodities into automotive fuel, either ethanol or biodiesel. In effect, the price of oil becomes the support price for food commodities. Whenever the food value of a commodity drops below its fuel value, the market will convert it into fuel.[131]

The results of these trends, Brown concludes, will have a devastating effect on the world's poorest populations:

> For the 2 billion poorest people in the world, many of whom spend half or more of their income on food, rising grain prices can quickly become life threatening. The broader risk is that rising food prices could spread hunger and generate political instability in low-income countries that import grain, such as Indonesia, Egypt, Nigeria, and Mexico. This instability could in turn disrupt global economic progress. If ethanol distillery demand for grain continues its explosive growth, driving grain prices to dangerous highs, the U.S. government may have to intervene in the unfolding global conflict over food between affluent motorists and low-income consumers. [132]

The controversy that exists here is precisely because resources are limited. There is only so much grain. The resource cannot fulfill every demand. Will this resource provide fuel for the energy guzzling SUV's of wealthy America, or will it be used to feed the hungry children of Ethiopia?

How would this issue be addressed by our stipulated freedom? First, there is freedom. To manipulate the nature of corn to produce energy for our human made machinery is God's extended freedom to humankind. We must ensure, however, that this action of biotechnology does not violate God's core values of love and justice. What would justice require? It would require a balance of resources among the world's population. There must be grain available for food, as well as for fuel. We could stop there, if there were enough grain for all. In a world of limited resources, however, we have to consider the second core value. What would love require? Agape love is expressed in self sacrifice for the needs of others. In God's economy, fuel for an SUV is not as important as food for a starving child in an impoverished nation. The core value of love would require the biotechnology of ethanol production to be purposely restricted in order to preserve enough grain for the food supply. The fuel dependent lifestyles of the world's rich would have to accept a level of sacrifice in order to meet the needs of the world's poor. That is the essence of Agape love. It is a stipulation on our permission of God to employ powers over nature. If we move full speed ahead in consuming the world's resources of grains for the conversion into fuels, we are utilizing God's power without God's purposes. We would be taking God's name in vain. Love is an important stipulation for us to remember as we travel down the road of biotechnology.

We have observed the core values of God: love and justice. With this information in hand, we are ready to proceed with the goal of this chapter.

On to the Compass

In order to exercise the freedom over nature that God has given us in a responsible way, we must travel down the road of possibility with a compass in hand. This compass would point us toward the direction of God's permission. In visualizing this direction, let us settle on the course of "due North". In doing this, there is no intent to imply any actual value to the physical realities of direction. The articulation of directions is merely metaphorical.

If due north is the direction of God's will, then we can find that spot on a typical compass, and exchange the word "North" with our two core values, love and justice (fig. 1).

Love & Justice

W   E

S

Figure 1

If a certain action of biotechnology, in either its production or implementation, shows evidence of moving in the direction of God's core values, the compass would show that this action is in line with God's stipulated freedom. Once again, let us be reminded that it doesn't matter how "extreme" or how beyond the scope of the natural such an action is. If it aligned with the direction of love and justice, creating the world's first humanzee could conceivably be within God's stipulated freedom.

In such a compass, there are other polarities that are off the mark. In figure 1, these remain as they are on a typical compass: East, West, and South. It is clear by looking at the compass that some polarities are closer to the desired direction than others. East, for example, is closer than South, which is, in fact, in the totally opposite direction. As we develop the compass to guide us, we will need to replace these words with appropriate concepts. The easiest one to tackle would be the replacement for the word "South". What is the opposite of justice? What is the opposite of love? If justice is the balance of resources among the world's population, then its opposite would be the growth of imbalance. If Agape love is the selfless concern for others, its opposite is, simply, selfishness (fig. 2).

Love & Justice

W   E

Increased Imbalance
Selfishness

Figure 2

Included in this "due South" position would be all manipulations of nature that have as their purposes to hurt or destroy.

Having established the polarities that directly lead into the path of our stipulated freedom and those that lead in the directly opposite direction, we are left with the two remaining markers. These would represent actions of biotechnology that are not clearly opposed to God's values, but don't quite hit the mark either. They would still amount to "taking God's name in vain", but with some intentionality, they might be moved into the more "due North" direction. There also exists the possibility that they could just as easily move more "South". On one side, we can place those actions of biotechnology that seem to been neutral in relation to God's core values. These are actions that don't promote equality in the world, but don't seem to add any strength to the imbalance either. On the other side, we can address the element of risk. Here we could place actions of biotechnology that have noble goals, but have the potential for unintended consequences that could bring hurt to others, or could exasperate the social imbalance of the world. As in the case of neutral actions, these actions too can move either up or down with some applied intentionality. With these concepts added, our compass (fig. 3) is complete:

Love & Justice

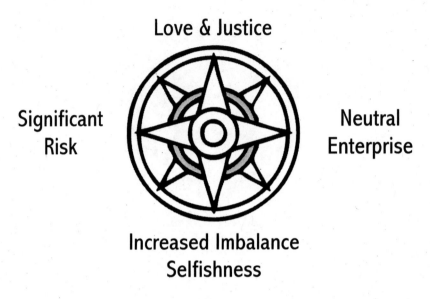

Significant Risk

Neutral Enterprise

Increased Imbalance Selfishness

Figure 3

This compass is offered as a tool to evaluate various biotechnological actions in relationship to God's stipulated freedom to exercise creative powers over nature as long as the core values of love and justice are preserved. As a tool, it is admittedly quite basic and rudimentary. It is unable to answer all of the ethical issues surrounding biotechnology, to be sure. It can be a helpful guide, however, in doing some initial measuring of the moral consequences of human activity in the realm of biotechnology. Perhaps this tool, with its supporting theology, could be helpful to individual people and communities of faith that are wrestling with their own convictions about the new frontiers of biotechnological activity. It is my hope that these individuals and religious communities will use this compass in their own thoughts and discussions about ethics and biotechnology. In an effort to demonstrate how the compass could be applied, the next several chapters will address specific broad areas of the biotech frontier. *The limits of this book do not allow a full treatment of these issues.* Each one could be a thesis in itself. In the following chapters, we will gain an introductory knowledge of these biotech frontiers, and then evaluate their direction in relationship to the compass. Some of these issues seem more feasible than others. Some are clearly on the fringes of possibility as we stand here at the beginning of the 21st century. All of them, however, would have seemed as mythical as the ancient Greek gods just a generation ago. It is fitting, then, that we allow the characters of Greek mythology to introduce each subject. Let us, therefore, move forward to consider a biotech pantheon with our compass in hand.

Chapter 3 Notes

[99] http://www.rotary.org/aboutrotary/4way.html

[100] Philippians 4:8

[101] Jack Nelson-Pallmeyer is the Professor of Justice and Peace Studies at the University of St. Thomas in St. Paul, Minnesota.

[102] Obery Hendricks Jr. is the Professor of Biblical Interpretation at the New York Theological Seminary and visiting scholar at Princeton Theological Seminary.

[103] Nelson-Pallmeyer, Jesus Against Christianity. Reclaiming the Missing Jesus., p.1.

[104] Nelson-Pallmeyer, p. 5

[105] Ibid, p. 10.

[106] Hendricks, Obery M. Jr. The Politics of Jesus. *Rediscovering the True Revolutionary Nature of Jesus' Teachings and How They Have Been Corrupted*, p. 14.

[107] Hendricks, Obery M. Jr., p. 43-44.

[108] Ibid, p. 44.

[109] Brueggemann, Walter. Old Testament Theology. *Essays on Structure, Theme, and Text*, p. 84

[110] Brueggemann, Walter. To Act Justly, Love Tenderly, Walk Humbly: *An Agenda for Ministers*, p. 5

[111] Gottwald, Norman K. The Hebrew Bible. *A Socio-Literary Introduction*, p.357

[112] Hendricks, Obery M. Jr. The Politics of Jesus. *Rediscovering the True Revolutionary Nature of Jesus' Teachings and How They Have Been Corrupted*, p. 14.

[113] http://www.query.nytimes.com/gst/ fullpage.html?sec=health&res=9C04E1DA1238F932A05756C0

[114] Mark 2: 23-27

[115] Goppelt, Leonhard. Theology of the New Testament. Volume 1: *The Ministry of Jesus in its Theological Significance*, p. 94

[116] I Corinthians 10:23-33

[117] Luke 4:18-19

[118] James 5:1-6

[119] Matthew 22:34-40

---

[120] Chilton, Bruce and McDonald, J.I.H. <u>Jesus and the Ethics of the Kingdom.</u>, p. 92
[121] Schnackenburg, Rudolf. <u>The Moral Teaching of the New Testament</u>, p. 328-329
[122] I Corinthians 13:13
[123] Collins, Raymond F. Christian Morality. *Biblical Foundations*, p. 137
[124] Collins, p.138
[125] en.wikipedia.org
[126] http://en.wikipedia.org/wiki/Greek_words_for_love
[127] Lewis, CS. <u>The Four Loves</u>, p. 146
[128] Ibid, p. 16
[129] http://www.cbsnews.com/stories/2007/01/23/ politics/main2391957_page3.shtml
[130] http://www.earth-policy.org/Updates/2006/Update55.htm
[131] Ibid
[132] Ibid

Enter Pegasus

The Compass Applied to Plant and Animal Biotechnology

Pegasus is no normal horse. He seems to be genetically modified. Pegasus is a chimeric of sorts, in that he seems to be composed of genetic parts from different sources than what normal nature has provided. So far, biotechnology has not yet produced a flying horse. The accomplishments in genetic manipulations in plants and animals, however, have been significant and impressive. Given what has already transpired, a Pegasus-like creature is not so difficult to imagine anymore. Just what is possible through animal and plant genetic modification? A few examples, out of a large body of evidence, may illustrate.

In 2004, researchers from the Japan International Research Center for Agricultural Sciences, working with scientists from the International Maize and Wheat Improvement Center of Mexico, successfully inserted the gene DREB1A from the plant Arabidopsis thaliana into wheat. This gene has been proven to allow its natural host, which is a wild mustard plant, to be extremely drought resistant. Studies are now underway to see if the same drought resistance capability exists in the genetically modified wheat. If so, then many countries currently facing famine and economic collapse could find a viable savior in drought resistant wheat.[133]

The cheetah is one of the most endangered animals in the world. Without a major turn-around, the cheetah is likely to go extinct within the next 15 years. One of the recognized problems with the cheetah is its lack of genetic diversity. The dominant theory is that the cheetah neared extinction during the last ice age, and the modern cheetah has emerged from a very limited gene pool. If the cheetah is to be saved from extinction, genetic diversity, not provided by nature, must be introduced into the species. These efforts are currently underway in various labs and zoos around the world.[134]

In the December 31, 2006 report published in *Nature Biotechnology*, the Hematech subsidiary of the Kirin Brewery Company announced that it had successfully produced healthy "prion protein-knockout cows."[135] The prion protein can become misfolded into infectious particles that lead to bovine spongiform encephalopathy in cows (mad cows disease) and, if passed on to humans through meat consumption, can lead to a lethal disease known as Creutzfeldt-Jakob disease. The research team believes that this accomplishment could pave the way to a safer food supply of prion protein free products.

One could add to these examples "ad infinitum". There are countless examples of genetically modified plants and animals, with an endless assortment of benefits provided for both humans and nature itself. Notwithstanding these benefits, there is a degree of controversy surrounding the biotechnology of plants and animals. Some of the objections would be dismissed forthright in this book. There are those who would object to the genetic manipulation of plants and animals because they believe that humans should not meddle with God's creation. In other words, they see nature as sacred ground. We have already tilled that ground, and have asserted that God has given humanity a freedom to manipulate nature within the stipulations addressed. Those stipulations are about values, not physical nature. Humans may, indeed, meddle with God's creation.

How would the genetic manipulation of plants and animals measure up under God's stipulated freedom? If we put this subject to the test of the magnetic pull of love and justice, how does it measure up? Likewise, what avenues of biotechnology involving plants and animals would be shown to be unaligned with God's values, and thus taking God's name in vain? Let us hold out the compass, and see which way Pegasus is directed. We will begin this evaluation with the issue of GM plants, and then move forward to the consideration of animals.

GM Plants Aligned to Love and Justice

Movement toward Love and Justice

The genetic modification of plants can, in so many ways, align perfectly with the direction of love and justice. The ability to make food products safer and more nutritious for human consumption is a wonderful way of protecting the people of the earth from disease. The resources of time, expertise, and finances poured into these areas are often sacrificial on the part of advantaged nations for the sake of poorer ones, thus perfectly aligned with the goals of God's love. God's value of justice, in which the resources of the world are fairly distributed among the inhabitants of the world, can be even better served through biotechnology than in the natural order. In many wealthy areas of the world, diets are diverse and rich in vitamins and iron. In many poor areas of the world, diets are limited to a few staples with limited nutritional value, such as rice. Recent advancements in biotechnology have led to the creation of fortified rice that can add new levels of nutrition to some of the world's poorest regions. In the online journal of Plant Physiology, Mary Lou Guerinot and David E. Salt, from the Department of Biological Sciences at Dartmouth College, wrote:

> Transgenic rice plants expressing the soybean ferritin gene contained three times as much iron in its seeds as untransformed plants. As one-half of the world eats rice everyday, genetically engineered rice with higher levels of ferritin and lower levels of phytic acid, which impedes iron absorption, would be a significant achievement.[136]

Indeed, biotechnology is enabling science to meet the basic needs of nutrition in new and bold ways. As we have already observed, biotechnology is also enabling countries with less viable land and water resources to grow crops that have, until now, been impossible to grow. This would have an impact not only on the health of the citizens of these nations, but the economy of them as well. This may allow the world's economic resources to be more evenly distributed. In many ways, God's values of love and justice can abound through the genetic engineering of the world's food supplies. By and large, human activity in GM plants seems to be within God's stipulated freedom. Given this positive position on the biotechnology of plants, however, there are certain aspects that would be called into question when put to the test of the compass.

Movement toward Significant Risk

Although much good can come from bioengineering processes with vegetation, there are ways in which this activity could begin to move off the desired course and into the direction of potential and significant risk. One such risk was observed by researchers at Cornell University in a 1999 study. The study concluded that genetically modified corn that had a built in resistance to a corn-boring caterpillar was inadvertently killing off the already endangered Monarch Butterfly.[137] This study caused a large uproar, until several other scientific authorities showed the research to be faulty. Even so, the debate about Bt Corn (as it is known) continues. The issue highlights one of the possible ways in which the bioengineering of vegetation can carry a significant risk. Genetic manipulations of one crop species can inadvertently cause harm to other species, both plant and otherwise.

A related risk is the cross pollination that can occur between genetically modified plants and natural ones. In a report released in December of 2002, the National Institute of Ecology announced that transgenic genes from one crop of modified corn had cross pollinated with crops of non modified corn in several towns that were surprisingly distant from the source location.[138] What risk this presents is, as yet, unclear. What is clear is that those who are taking the risks are putting others at risk. In this, biotechnology does not seem to be expressing a selfless concern for others. If not, is it consistent with God's core value of Agape love?

A final risk to be presented here is that of "superpests". In our own bodies, we are well aware of the constant arms race that occurs between antibiotics and germs. Stronger antibiotics eventually lead to stronger germs. This same evolutionary process can occur between plants that are genetically modified with pesticides and the insects that normally feed on them. Although the biotech industry could continue to combat superpests with superplants, plants that do not have such engineered protection will be virtually defenseless against the pests.[139] Once again, the risk taken here is putting others at the greater risk.

These risks, along with others, begin to shift the arrow of our compass away from the "due North" position of preserving God's values of love and justice. As we have seen, however, there are abundant ways in which the bioengineering of plants aligns perfectly with those values. The proper exercise of this technology, then, should be to evaluate risk factors thoroughly and aggressively. If the levels of risk are deemed small, or can be mitigated, they should be. If risks are deemed real and significant, then the cautionary yellow light of God's stipulated freedom may indicate a need to stop or slow down. Such responsible evaluation of risk is, in fact, happening. In 2000, the Biotechnology and Biological Sciences Research Council (BBSRC) and the Natural Environment Research Council (NERC) in Britain embarked on a multimillion dollar study that addresses these issues. The National Environment Research Council reported that:

> Many of the projects are directly relevant to risk assessment issues and public concerns about GM crops and will extend existing areas of research. For example, one project will investigate the interactions between GM plants, pests that feed on them and parasites that feed on the pests, with a view to evaluating the risks that "superpests" might inadvertently be produced by some GM crops. Others will extend research into the frequency and significance of gene transfer between microbes in the soil, and between plants and microbes.[140]

When science takes the issue of risk seriously, and does the hard and important work of assessing that risk with the intention to follow the recommendations of the data even if it means curtailing biotechnology, the element of risk is being addressed in a responsible way. In this dynamic between pursuing GM plants to better the lives of the world's populations and the addressing of potential risk, the arrow on our compass is experiencing two different magnetic pulls. The potential for love and justice is strong, and the presence of significant risk is real. At this point, the compass would show Pegasus running in a "northwest" direction, somewhere between the fulfillment of God's core values (i.e. righteousness) and the possible unintended harm to others and/or the environment. Risk, however, is not the only threat to a due North course. In GM plants, there is a pull south as well.

Movement toward Increased Imbalance or Selfishness

The global human community is deeply infected with injustice. The Helsinki-based World Institute for Development Economics Research of the United Nations University (UNU-WIDER) has recently released a comprehensive study of inequality factors across the world. This study showed that "the richest two percent own half of the world's wealth".[141] To further elaborate the findings, the report states:

> The most comprehensive study of personal wealth ever undertaken also reports that the richest 1% of adults alone owned 40% of global assets in the year 2000, and that the richest 10% of adults accounted for 85% of the world total. In contrast, the bottom half of the world adult population owned barely 1% of global wealth.[142]

Given such deeply imbedded inequality throughout every dimension of human life, it is nearly impossible for any engineering enterprise to be in full compliance with God's core values of love and justice. In every issue we will discuss, one could raise the global trump card of economic inequality. Wonderful medical breakthroughs fall into the net of global economic inequality, and are therefore of no help to the poor or uninsured. New breakthrough drugs to offer hope to people living with HIV don't seem to find their way into the parts of the world that are most devastated by the virus. In an abundance of ways, and for a long time now, the realities of injustice and selfishness have reigned supreme. Overall, the human community is not doing a very good job of following God's stipulated freedom.

Given this reality, any issue of biotechnology will fall into already demarked territories of global injustice. As we evaluate these particular areas, we must acknowledge that as a given. In putting these issues to the test of the compass, our question must move forward from that basic reality. There is already imbalance in the world. Will this enterprise increase that imbalance? There is already a deeply imbedded selfishness in this world. Will this enterprise foster even more, or does it shine forth a quality of self sacrifice for the needs of others as a kernel of Agape love?

Perhaps more than many other potentialities of biotechnology, the genetic modification of plants has a very strong potential for doing the work of love and justice in the world. We have already seen a few examples. One could cite many more, almost ad infinitum. The pull toward love and justice is very strong. Not withstanding the global climate of inequality we have already observed, are there particular ways in which GM vegetation technologies could deepen that inequality? Is the enterprise susceptible to selfishness and greed?

In the last chapter, we witnessed one such example of that happening. It is the competing commodities of food and fuel. Wealthy nations have the larger fuel needs. Poorer nations have the larger food needs. In the growing arena of ethanol production, as we have seen, the limited resource of global vegetation is, more and more, being devoted to fuel production. This is largely driven by two factors: The desire of powerful nations to have a new fuel source to maintain a way of life without foreign dependency, as President Bush stated in his 2007 State of the Union Address; and the economic advantage that large corporate farms have in channeling their crops toward fuel production rather than food production. In this example we find the two things antithetical to God's core values. Injustice flows as the imbalance of equality is exasperated, and those who stand to benefit from this technology are reaping selfish rewards rather than doing the selfless things for the sake of the other.

In this poignant example, we can witness ways in which the biotechnology of plants can move in the opposite direction of God's core values. The magnetic pull away from God's stipulated freedom is strong here. Once we recognize the possibility of a scenario such as this, it is important to seek ways in which this impending injustice is adverted, or at least mitigated. As we observed earlier, President Bush encouraged a growing use of ethanol "using everything from wood chips, to grasses to agricultural wastes."[143] Focusing research on ways in which non-edible sources of vegetation can be devoted to ethanol production, allowing the edible parts to meet the growing needs for food throughout the world, could be a wonderful way of mitigating the possible injustice here. Some would say that these sources are not the best for the production of ethanol. If we went in this direction, we would be sacrificing some things for the sake of preserving food sources for others. That is precisely true, and it is precisely the nature of Agape love.

In this brief consideration of the issues related to the bioengineering of plants and vegetation, we have seen both the positives and negatives of this biotechnological enterprise. If a careful and responsible risk assessment program is continually engaged; and if agencies of influence seek ways in which the needs of the world's most vulnerable citizens are protected, even if it means curtailing the total potential of certain endeavors; then the predominant influence of good that genetically modified vegetation can offer pulls the directional arrow strongly toward the due North position. This aspect of biotechnology is headed in the general direction of God's stipulated freedom. There remains the possibility, however, that unintended negative effects will happen. There remains the possibility that actions will, in fact, selfishly exasperate the imbalance of equality in the world. Factoring these possibilities into the journey, our compass would still show some magnetic pull toward risk and injustice. A generalized compass reading of the biotechnology of plants, then, would look something like figure 4:

Genetic Modification of Plants

Love & Justice

Significant Risk

Neutral Enterprise

Increased Imbalance Selfishness

Figure 4

One should not be discouraged to see that the arrow doesn't show 100% alignment with God's core values. A total alignment would represent perfection, which no human being could claim. Likewise, a total alignment would imply that there are no more questions to ask, and no more concerns to consider. The arrow shows a general movement toward God's core values. We should be ever diligent to keep the arrow moving ever farther north, and no farther south.

GM Animals Aligned to Love and Justice

Movements toward Love and Justice

Such is our compass reading of the bioengineering of plants, but Pegasus was not a plant. For many people, the ethical issues become more concerning when the subject of bioengineering moves from the realm of plant life to the realm of animal life. This is largely based on our internalized sense of graduated life, which places animals at a higher level than plants. Likewise, animals more like us (mammals, e.g.) are generally considered more sacred and protected than lower life forms (fish, e.g.) Many people become a bit uneasy when biotechnology puts animals in its sights.

Even so, the bioengineering of animals is in full swing. In many ways, the results of these efforts are leading to tangible benefits for the people of the world. To address the promising aspects that arise concerning animal bioengineering, let us focus on multifaceted issue of cloning.

Most people are familiar with Dolly the Sheep, the first publicly acknowledged cloned mammal. Since Dolly, a host of other animals have been cloned, including mules, rabbits, cats, pigs, cows, mice, and (a little too close for comfort for some) monkeys.[144] The science of cloning is advancing, thanks largely to the experiments with animals. Many people think that all cloning is the same. In reality, there are different cloning techniques, which have different goals and results.[145] *Recombinant DNA Cloning* has been used for many years. This process inserts a DNA sample under study into a self-replicating plasmid that fosters multiple copies of the DNA fragment. This technique has proven to be a significant help in molecular biology, leading to some wonderful advances in medical abilities.

Reproductive Cloning, which captures most of the headlines, utilizes a process known as "somatic cell nuclear transfer" to remove one piece of genetic material from the nucleus of a donor and insert it into an egg that has had its nucleus removed. The egg is then induced to begin development. Once it reaches a suitable stage, the egg is implanted into a host. The baby born has virtually the identical genome of the donor. Some of the genome of the baby, however, is determined by the mitochondria in the cytoplasm of the enucleated egg.

A third form of cloning, known as *Therapeutic Cloning*, is a process by which embryos are created for the purpose of harvesting stem cells. This form of cloning will be discussed in our chapter dedicated to stem cell research.

The growing abilities to mix and match genetic materials through cloning does not only allow for nearly identical copies of a donor. The same processes can allow different genetic material to be purposely engineered to produce certain traits within a new animal. This process of genetic modification can lead to many promising possibilities. Genetically modified pigs, for example, could become a seemingly limitless resource of possible organs that could be transplanted into ailing humans. This process, known as xenotransplantation, is already making significant process. Why pigs? The web site of the Human Genome Project, *Genomics.Energy.Gov*, explains:

Why pigs? Primates would be a closer match genetically to humans, but they are more difficult to clone and have a much lower rate of reproduction. Of the animal species that have been cloned successfully, pig tissues and organs are more similar to those of humans. To create a "knock-out" pig, scientists must inactivate the genes that cause the human immune system to reject an implanted pig organ. The genes are knocked out in individual cells, which are then used to create clones from which organs can be harvested. In 2002, a British biotechnology company reported that it was the first to produce "double knock-out" pigs that have been genetically engineered to lack both copies of a gene involved in transplant rejection. More research is needed to study the transplantation of organs from "knock-out" pigs to other animals.[146]

Xenotransplantation is not the only benefit of genetically modified animals. The possibilities abound. As the BBC announced on August 21, 2000, two goats by the names of Webster and Pete were, in fact, interesting clones. During the cloning process, these goats were given a spider gene. The hopes were that these goats would produce a type of "silk milk", a super strong yet bio-based material stronger than steel, yet compatible with the human body. This material, known as *biosteel,* offers great hope for use, according to Nexia, the company that produced the goats. The BBC report states:

> BioSteel could be used for strong, tough artificial tendons, ligaments and limbs. The new material could also be used to help tissue repair, wound healing and to create super-thin, biodegradable sutures for eye- or neurosurgery. "The medical need for super-strong, flexible and biodegradable materials is large," said Costas Karatzas, Nexia's Vice President of Research and Development.[147]

From organs to super biosteel, GM animals are a revolutionary resource. Add to this the efforts of producing cows that produce milk high in protein or other vitamins, or perhaps carrying antibiotics to tackle childhood diseases; or the prion protein free cows that are immune to Mad Cow Disease; or any number of possible benefits that could be listed here, and the evidence becomes clear: the genetic modification of animals is benefiting human life in remarkable ways. Given this, how would this issue line up to the compass? In introducing this issue, we have already seen the evidence that would show the many ways in which GM animals could promote God's justice and love in the world. Diseases can be conquered; food sources can be made safer and more nutritious; endangered animals can be rescued; and animal research can lead to tremendous human applications. Does the arrow, therefore, align perfectly with God's core values of love and justice? Actually, even though much good can come from these technologies, the answer would have to be no. Let us now put this issue to the compass, and see what divergent paths emerge.

Movement toward Neutral Enterprises

The ability to manipulate the very forms of life, and create new life forms by either significantly modifying the genetic makeup of a known species, or merging the genetic materials of two distinct species in order to create a hybrid, can be an intoxicating power. For most activity in the world of animal biotechnology, actions are taken with strong therapeutic aims in mind. In an earlier part of this book, we consider what is, for many people, a controversial hybrid between humans and mice. The Stanford University's Institute of Cancer/Stem Cell Biology and Medicine, under the direction of Dr. Irving Weissman, has created mice with fully human immune systems, as well as significant human brain structure.[148] There are those who object strongly to this research because they believe that it is crossing a sacred line of nature that should not be crossed by human beings. In light of the stipulated freedom presented in this book, the actual gene manipulations that have led to these mice are not the problem. The real question here is, are these manipulations pursuing the cause of God's core values? In this case, the research is being done with the primary goal of discovering breakthrough treatments in diseases such as Parkinson's disease. The arrow on our compass would move north.

There are other hybrid goals, however, that don't seem to have such clear therapeutic ends in mind. Early, we considered the case of Dr. Stuart A. Newman, professor of cell biology and anatomy at New York Medical College, who attempted to secure a patent for a "humanzee."[149]. As you recall, his stated purpose for this patent was to keep others from attempting to create a human-chimp hybrid for mere "curiosity" sake. It would be a safe bet to believe that somewhere in the world today, someone with the needed resources is attempting that very thing. The goal is not evil, per se; but it is also void of any motive for love and justice. Such research, if it exists, is truly utilizing God's power without God's purpose in mind. In terms of God's core values, it is a neutral enterprise.

As biotechnology advances, and more and more possibilities emerge, the attraction toward novelty and of doing something first can easily become the main impulse for action. If we wish to exercise our freedom to manipulate nature in a responsible way, we must find the ways to self-impose limits to our own abilities and powers. Exercising those powers in ways that are neutral to God's core values is off the mark.

Movement toward Increased Imbalance or toward Selfishness

Even though there is a potential that the powers to create new creatures by cloning techniques will be driven by neutral purposes driven by curiosity, this activity would pale in comparison to the overwhelming amount of animal bioengineering that would be for good and therapeutic purposes. In applying the compass to this path, the magnetic pull  toward neutral enterprises would be weak in comparison to the pull of God's core values. If this were the only issue of concern in the bioengineering of animals, the compass would be quite reassuring. With the most significant pull away from God's core values being actions of injustice and selfishness, however, the compass may begin to show otherwise.

There are, of course, questions of financial benefit to those who create GM animals, especially if they are able to patent their new creations. In addition, as we have already observed, breakthrough medical therapies derived from the bioengineering of animals can fall into the wide net of inequality that determines who can have access to those therapies, and who can not. Beyond these obvious possibilities of injustice and selfishness, there is an issue that could drastically challenge the whole enterprise of animal bioengineering within the context of God's stipulated freedom. At the heart of justice, there is the protection of the world's most vulnerable. At the heart of Agape love, there is self sacrifice for the sake of "the other." Are not animals among the most vulnerable in this world? Is it possible that animals may, in fact, be "an other" to humanity? To put the question in Martin Buber's words, are animals the "thou" to our "I"?

In a worldview that sees humanity as a separate and special creation, placed in nature yet apart from nature, it may be easy to see a clear dividing line between the ethics that govern our treatment of people and the ethics that govern our treatment of animals. Seen in this way, animal life becomes, per Heidegger, our "standing reserve"; value-free materials to be used for our purposes.[150] The advancing sciences of NeoDarwinism, however, have made the point abundantly clear: Humanity is not *apart* from nature. Humanity is a *part* of nature. That which separates us from the animal kingdom is really quite minimal. We are made of the same stuff, virtually the same DNA, with far more genetic similarities than could have been earlier imagined. We may not be surprised to see how genetically close we are to a chimpanzee, but our proximity to something as seemingly foreign as a mouse has been demonstrated by molecular science. In his book The Language of God, Francis Collins (Head of the Human Genome Project) writes:

> The overall size of the two genomes (human and mouse) is roughly the same, and the inventory of protein-coding genes is remarkably similar. But other unmistakable signs of a common ancestor quickly appear when one looks at the details. For instance, the order of genes along the human and the mouse chromosomes is generally maintained over substantial stretches of DNA. Thus, if I find human genes A,B, and C in that order, I am likely to find that the mouse has counterparts of A,B, and C also placed in the same order, although the spacing between the genes may have varied a bit. In some instances, this correlation extends over substantial distances.[151]

Collins goes on to anticipate the argument that God simply used the same general pattern and principles in creating the mouse and the human, yet they were different creations. He counters this by discussing the discovery of ancient repetitive elements (AREs) known as "jumping genes". These genes apparently copy and insert themselves in a useless place within the genome. Some of these jumping genes end up becoming damaged in the jumping process, thus ensuring that they have no practical use to the genome. They are, truly, "junk DNA". Yet, according to Collins, " when one aligns sections of the human and mouse genomes, anchored by the appearance of gene counterparts that occur in the same order, one can usually also identify AREs in approximately the same location in these two genomes.[152] The conclusion reached by this analysis is that mice and humans share a common ancestor, one not so far back in earth time.

We have a true natural relationship, then, with animal life. Humans have experienced an evolutionary pull toward larger and complex brains. Along the way, we somehow acquired self awareness and even spirituality. Yet we share common ancestors with the animals around us. They are not just the stuff of a world that we inhabit. They are our "other".

If this is the case, then it would raise a host of questions concerning the biotechnology of animals. Are we following God's core values of love and justice when we submit animals to research? What does this do to our human diets? Should we become vegetarians, or strive to develop synthetic foods to replace our natural diets, much like we did when we developed Polyester to replace animal furs? How far one would want to take the protection of animals and the elevation of animal rights is a very fluid conversation, in which different people would end up in many possible positions along a spectrum of answers. It is interesting to note, however, that some countries are beginning to take some definite steps toward animal rights. In the May 17, 2002 edition of the *New York Times*, journalist Ruth Gledhill reported that:

> Germany has become the first European nation to vote to guarantee animal rights in its constitution. A majority of lawmakers in the Bundestag voted on Friday to add "and animals" to a clause that obliges the state to respect and protect the dignity of humans. The main impact of the measure will be to restrict the use of animals in experiments.
>
> In the end 543 lawmakers in Germany's lower house of parliament voted in favor of giving animals constitutional rights. Nineteen voted against it and 15 abstained.

Where the line is to be drawn here is a difficult question. Obviously, there is a food chain in this world of God's creation. Killing animals for meat seems to be an inherently natural thing to do. Killing animals for clothing or for medical research is one step beyond. Where it all falls in God's intention for love and justice is not easy to determine. It does become easier, however, when animal life is expended not for the basic needs of humans, but to compensate for human indulgences and selfish lifestyles. Wearing a bear fur because it is the only way to keep warm for an ancient Eskimo is not in the same category as wearing a mink fur to demonstrate one's class and sophistication. Such activity may violate God's core values of love and justice. Wearing such a mink fur may be taking God's name in vain.

If this is the case, then how should we evaluate the bioengineering of animals? Suppose we do find the bioengineering powers to create donor pigs that are actually living hosts to human-compatible livers and hearts. Although many things can lead to liver and heart disease, medical science has concluded that the major causes are lifestyle issues. We overindulge in sugars, fats, and alcohol. If we genetically alter animals so that their organs can replace our own when we have mistreated them by our overindulgent lifestyles, is this not, at the root, selfish behavior? Is not Agape love violated? Has not justice been withheld for the world's most vulnerable?

How the bioengineering of animals aligns with the compass, then, hangs on a major question. Are animals included in God's expectation of love and justice? Answer this question with a negation, and the compass would swing toward the north. Answer it affirmatively, however, the compass would swing strongly to the south. The differences are so significant, that it would be hard to demonstrate an overall compass view that would allow for either possibility. To summarize an overall evaluation of the genetic modification of animals, two possible compasses are offered.

The genetic modification of animals can offer much positive advancement for human life. Fortified food products, new bio-friendly materials, and new resources for organ transplants could extend love and justice far and wide. Mitigating this strong pull toward God's core values, however, are the possibilities of neutral enterprises driven by curiosity or competition, and greedy efforts by some who wish to design patents and control creations. *If animals are excluded from the rights and values God placed on humans*, Our compass would look something like figure 5:

Love & Justice

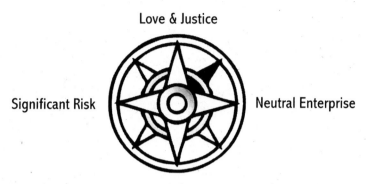

Significant Risk   **Neutral Enterprise**

Increased Imbalance Selfishness

Figure 5

Given everything that was said in the preceding paragraph, we must now entertain the possibility that *animals are included in the rights and values God placed on humans*. If this is the case, then much of the realm of animal bioengineering would be off course. The compass would reflect this: (fig. 6)

Love & Justice

Significant Risk   **Neutral Enterprise**

Increased Imbalance Selfishness

Figure 6

Obviously, a major question hangs in the balance. Although it is beyond the scope and potential of this study, much theological attention will need to be given to the status of animals in God's economy of love and justice.

Chapter Conclusions

In this chapter, we have exercised the use of the compass to evaluate a major section of the bioengineering world. It is my hope that the reader will continue to engage the issues of plant and animal bioengineering, utilizing the compass as the evaluation tool. As this chapter illustrates, the application of God's stipulated freedom passes no judgment on the actual manipulations of nature. It begins with the fundamental freedom that we have to manipulate nature. As this chapter has demonstrated, however, the stipulated freedom is far from a totally permissive ethical stance. Physical nature is not sacred; love and justice, however, are. These core values provide a very challenging goal for the exercise of biotechnology in accordance with God's stipulated freedom. When God said "You shall not take my name in vain", God gave us a challenging word.

Now that we have practiced the use of this compass on plant and animal life, it is time for us to switch our attention to human life. The same technologies that are being employed at the animal level can, albeit in a more complicated way, work on humans as well. What guidance does the compass offer to the emerging fields of human biotechnologies? In the next chapter, two major issues will be put to the test of the compass. The Greek gods Hermes and Hercules will lead the way.

Chapter 4 Notes

[133] http://www. Seedquest.com

[134] BioScience, vol. 41, No. 9

[135] Hematech new release, 1/3/2007

[136] http://www. Plantphsysiol.org

[137] http://www.biotech-info.net/gemonarchs.html

[138] http://www.cqs.com/gmocorn.htm

[139] http://www.aboutbioscience.org/pdfs/GM_Foods.pdf

[140] http://www.nerc.ac.uk/press/releases/2000/22-geneflow.asp

[141] http://www.globalpolicy.org/socecon/inequal/2006/1206unustudy.htm

[142] Ibid

[143] http://www.cbsnews.com/stories/2007/01/23/politics/main2391957_page3.shtml

[144] http://www.ornl.gov/sci/techresources/Human_Genome/elsi/cloning.shtml#animals

[145] This information was gleaned from the "Cloning Fact Sheet" of the Human Genome Project, and can be found at the web site: http://www.ornl.gov/sci/techresources/Human_Genome/elsi/cloning

[146] Ibid

[147] news.bbc.co.uk/1/hi/sci/tech/889951.stm

[148] See pp 31-32 for the discussion and citations of this issue.

[149] See pp 92-93 for the discussion and citation of this issues

[150] See page 24

[151] Collins, Francis. The Language of God. *A Scientist Presents Evidence for Belief*, p. 134

[152] Ibid, p. 136

Enter Hermes and Hercules

The Compass Applied to Stem Cell Research and Prenatal Engineering

Stem Cell Research

It seems that out of all the biotechnology issues causing controversies in society today, stem cell research is at the forefront. In this issue, medical biotechnology is in a strong confrontation with both religion and politics. With so much controversy swirling around the issue, perhaps we could silence the noise of debate and seek, as objectively as we can, to apply the compass to see if the direction of stem cell research is within God's stipulated freedom.

Hermes has been chosen as the symbol of this issue, primarily because of the stories of Greek mythology related to his day of birth. Hardly out of the womb, Hermes demonstrated amazing powers and abilities. He traveled tremendous distances, corralled a herd of cattle, and invented the lyre. To see such power within such a premature life was truly the fodder of Greek legend and mythology. In the field of medical science, however, other tremendous powers are being discovered in human biology far more premature than Hermes. One of the most promising fields of medical research is centered on something that is located within a human embryo that is days old: the embryonic stem cell. Before we consider this issue in the context of the compass, a brief introduction to the field of stem cells is in order.[153]

As early as the eighteenth century, biologists have been fascinated by the ability of some creatures to regenerate their bodily parts. The written records of Abraham Trembley, for example, demonstrate this eighteenth century biologist's fascination with the apparent immortal-like ability of the hydra to constantly regenerate itself. Although few creatures have the extreme regenerative ability of the hydra, there still are plenty of creatures that are able to grow back lost limbs or tails, from the star fish to the lizard to a variety of insects.

As time went on, researchers began to wonder: Do humans share this trait, even to a tiny degree? If so, there would be the possibility to harness this regenerative power in order to create endless medical benefits. The search for this regenerative agent in the human body was on.

A major step forward in this search occurred in the early 1950's. A strain of mice was produced, in which the mice carried a particularly nasty tumor. What was odd about this tumor was that inside it, a variety of cells were found. These cells ranged from bone cells, to blood cells, from cardiac cells to neuron cells. As "mouse strain 129" was examined, it was determined that the tumors actually contained an embryonic cell that was capable of producing the many specialized cells of the body. Inside an embryo, this cell went to work generating a variety of cells. Outside the embryo, this cell behaved erratically, and led to the tumors that were experienced by the unfortunate mice of strain 129. This mysterious embryonic cell seemed to be the key to human regeneration. But for a long time, it remained hidden. It wasn't until our own recent times, 1998 to be exact, that a pair of scientists (James Thomson and John Gearhart) announced to the world that they had successfully identified the mystery embryonic cell. The *stem cell* was introduced to the world.

Since that time, major advances in understanding the role and function of stem cells have occurred. Scientists have discovered that there are actually three levels of stem cells in the human body. The first type, most generative in nature, is called the *Totipotent Stem Cell*. These stem cells are able to produce the entire organism of a fetus, including all of the supporting tissue needed for gestation. The totipotent stem cell is found in the fertilized egg. Totipotent stem cells are a bit more "potent" than useful for therapeutic purposes, due to their use of creating gestational tissues. Stem cells that have all the abilities to generate tissues and organs within a human being, yet without the gestational functions, would be the ideal. Such stem cells do exist. This second type of stem cell, upon which the controversy rests, is known as the *Pluripotent Stem Cell*. This is the stem cell that holds such tremendous promise. It is the cell that can produce the many specialized cells that make up the human body. It is the pluripotent stem cell that can be the basis for a wide array of future regenerative therapies. Pluripotent stem cells, however, can only be found in early embryos or in the reproductive tracts of unborn fetuses. In order to harvest them, embryos or fetuses must be destroyed.

The third type of stem cell is above the controversy, and is being used today in promising new medical therapies. It is the *Multipotent Stem Cell*. Multipotent stem cells, also known as "adult stem cells", are found in many parts of the human body. Umbilical cord blood has been found to be rich in multipotent stem cells. In most cases, they can be harvested from donors without endangering life. They have the ability to regenerate within their own specialized realm. For instance, the blood stem cell has been identified, and is already becoming a breakthrough treatment in leukemia and other blood related illnesses.

But as promising as they are, multipotent stem cells are limited in their ability. The real medical prize is in the more powerful stem cell: the Pluripotent, or "embryonic" stem cell. We are still in the early stages of exploring the possibilities of embryonic stem cells. In order to advance this stem cell research, scientists will need to harvest these cells from embryos or fetuses. Many researchers, wanting to stay clear of the abortion debate, have avoided the use of fetus reproductive tissues. What the researchers want to do is to harvest stem cells from embryos within the first several days of development. Such embryos can be produced in the laboratory. Utilizing the same technologies that enable unfertile couples to have children (i.e. *in vitro fertilization*) could easily provide a storehouse of human embryos from which pluripotent stem cells could be harvested. Many religious and political leaders have seen this as creating human life in order to destroy it, and have deemed it unethical. Similarly, in the last chapter the subject of *therapeutic cloning* was discussed. Through cloning techniques, human embryos could be created for the expressed purpose of harvesting their stem cells. Once again, however, it seems to many that in this method, human life is created *in order* to destroy it. The conflict between those who endorse therapeutic cloning and those who oppose it can be witnessed in a recent article in the South London Press, by reporter Clare Casey:

The creator of Dolly the sheep is to join forces with a South London team of researchers to clone embryos of humans with motor neuron disease (MND).Professor Ian Wilmut and Camberwell-based Professor Christopher Shaw were this week given permission to carry out the research known as "therapeutic cloning" at King's College. Professor Wilmut, who created Dolly, said: "This is only the second license ever granted by the Human Fertilization and Embryology Authority to use this technique. We will clone embryos that have MND from patients who have the condition - purely for research purposes." The cloning of human embryos will be done in Scotland before the cells are studied at King's College's School of Medicine building in Denmark Hill, Camberwell. King's College researcher Professor Shaw said the technique could dramatically speed up the process of finding life-saving drugs. He said: "Motor neuron disease is a relentlessly progressive muscle-wasting disease that causes severe disability from the outset."We have spent 20 years looking for genes that cause MND and to date we have come up with just one gene. By using the stem cells of humans with the disease it will speed up the whole process dramatically. Angie Coster, spokeswoman for the Motor Neuron Disease Association of South Thames, said she welcomed the move She said: "This means we are a step closer to medical research that has the potential to revolutionize the future treatment of MND. " But Dr David Jones, a Catholic lecturer in bioethics, said it should not be allowed to go ahead. He said: "To create a human embryo for the purpose of destroying it is homicide. It is demeaning the life of an embryo and sends out a dangerous message for future experimentation. "My view - which is that of the Catholic Church - is that this is experimenting on human beings."[154]

There is an abundant source of embryos, however, that were not created for the expressed purpose of destroying them. They are the embryos from the widely accepted practice of in vitro fertilization.

Today, in vitro fertilization is offering new hope to thousands of couples who have been unable to bear children. In laboratory procedures, sperm and egg from the husband and wife (or other donors if needed) are united and nurtured to develop. The resulting embryos are nurtured in the laboratory, and then planted in the woman's uterus. Although only a few embryos are actually implanted, there are usually dozens of embryos created for the couple. The excess embryos are frozen for future use. But for the most part, they are not used. They remain frozen, until they are finally destroyed. It is estimated that there are currently 400,000 such frozen embryos destined for destruction. If embryos are going to be destroyed, and those same embryos contain embryonic stem cells that offer such promise for future medical therapies, should the stem cells be harvested from them before they are destroyed?

This is the current fault line of the stem cell debate. In this case, human embryos are not created in order to be destroyed. To the contrary, human embryos are going to be destroyed regardless of whether their stem cells are intact or not. Many stem cell research advocates claim that the issue here is akin to the practice of organ donation.[155] In this view, the person on the hospital table is going to die. Given that reality, the person's vital organs are harvested. As a society, we have generally accepted this, and even see it as admirable. Is not the same thing happening, the argument asks, when stem cells are harvested from embryos that are going to die? Others oppose this comparison, highlighting the fact that in the case of organ donation, there is no choice but death. The frozen embryos have a full potential for life if they were to be planted into a uterus.

There is, obviously, a lot to be sorted out in the ongoing debates over stem cell research. What do we find when we evaluate this issue in the light of God's stipulated freedom? The first conclusion we could draw is that the actual manipulations of stem cells, in whatever form and fashion they take, would be within our general freedom to exercise creative power over nature. If we can induce stem cells to behave in ways quite contrary to a natural progression, as is being attempted in some research that we will observe momentarily; if we can direct stem cells to generate bodily tissues and organs according to our desires; if we can create embryos that produce stem cells but are missing other necessary ingredients for life development (as we will observe momentarily); any of these manipulations of a physical object known as a stem cell are within God's stipulated freedom. We must continually remind ourselves that the stipulations were not about nature, but rather about God's core values of love and justice. Let us, then, apply our compass to the pathway of stem cell research, and see how it aligns with those core values.

The Compass Applied to Stem Cell Research

Movement toward Love and Justice

Even though there is tremendous debate swirling around the issue of stem cell research, it is overwhelmingly accepted that stem cells have a tremendous potential as a source for breakthrough medical therapies. No one can question the benefits already derived from the use of multipotent (adult) stem cells in treating certain blood disorders. Promising studies are currently underway with cardiac stem cells and neural stem cells, two other types of multipotent stem cells that are readily available for research and development. Given what has already been demonstrated through these cells, it seems unquestioned that pluripotent stem cells would be an almost unprecedented (to date) advancement in our abilities to treat illness and disease. Notwithstanding the contextual problems of justice that affect all medical work (i.e. access for everyone), there seems to be little doubt that stem cell research could fulfill God's core values in magnificent ways. The magnetic pull toward love and justice would be, for stem cell research, tremendously strong.

Movement toward Significant Risk

Like many medical adventures, early results are dampening the fire of enthusiasm a bit in the realm of stem cell research. Doctors in Seoul, Korea, for instance, have recently reported that they had to suspend human trials of cardiac stem cell therapies because the initial progress of stem cells in opening closed arteries was beginning to reverse quickly.[156] Here in the United States, the Food and Drug Administration has suspended research into cardiac stem cell therapy due to an unexpected problem with their research dogs. When cardiac stem cells where infused into the blood vessels of the heart, they caused tears, known as cardiac microinfarcts, in heart vessels.[157] In addition to these problems, there are still major hurdles to overcome in terms of immune acceptance of stem cell therapies.

The risks mentioned here are no doubt just the beginnings of risks that will be discovered as stem cell therapies become more and more utilized. There is, then, an element of risk that would pull the arrow of our compass toward the west. We can be sure, however, that risk assessment will be vigorous and ongoing. The two cases cited demonstrate this aptly. When a risk is discovered, progress is suspended; researchers go back to the drawing boards and seek solutions to eliminate that risk, or at least mitigate it to some acceptable level. The magnetic pull toward significant risk, then, is quite weak. Even weaker, would be the magnetic pull toward neutral enterprise. One could claim that Abraham Trembley was initially driven by curiosity, and perhaps that was so. In so many instances, technological and medical progress have been initiated by acts of mere curiosity. In constructing the compass, we did not assert that curiosity is opposed to God's core values. Neutral enterprise is not the south to God's north. Neutral enterprises are just that: they do nothing to either promote or diminish. Yet even if the initial research was driven by curiosity, stem cell research today seems universally driven by its promise for medical research. The arrow of the compass would feel virtually no pull to the east.

Given the amazing potential of stem cell research, the fairly weak pull of significant risk, and the almost non-existent pull of neutral enterprise, stem cell research would, at this point in our discussion, do extremely well in relation to the compass. Perhaps we could visualize it this way: (fig. 7)

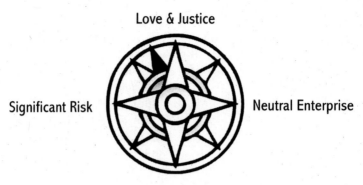

Figure 7

This may look very promising for stem cell research, but we have left the most difficult questions for last. As mentioned earlier, the current source of pluripotent stem cells is a living human embryo. The harvesting of stem cells from that embryo will destroy it. The hardest questions for us are these: When does protected human "life" begin? Does God's stipulation of love and justice protect an embryo that is only a few days old?

The challenge in finding a universally accepted answer to the question of the onset of "life" is that the question is beyond the scope of scientific analysis. Science can certainly explain the detailed mechanics of embryonic development. Even so, however, the label of "life" (as a label of individuated human identity) is a subjective assignment. According to one view, the earliest that one could identify as the beginning of life would be on day 14, when gastrulation (the organization of an embryo) occurs. The days prior to that are considered to be "pre-embryo," and therefore not definable as human "life." Typically, embryonic stem cells are harvested around the 8th day. As a pre-embryo, the collection of cells would not constitute a human life, but would represent what we might call a *life-potential*. If it is merely an element that has the potential to foster embryonic life, then is it special in any way? Cloning has demonstrated that one could take a cell from virtually anywhere in the body and use it as the catalyst for embryonic development. Dolly was cloned with a cell from an udder. No one would claim that the skin on my finger is sacred and protected by God's values of love and justice;

is an eight day old fertilized egg any different? Both, under the right circumstances, have the potential for fostering a new life.

Others, of course, don't accept this concept of a period of time in which the object under question is not an embryo but rather a "life-potential pre-embryo". For many, life begins at the moment of *conception*. Those who hold this view, however, will need to continually refine their understanding and definition of conception. We have left the time when human reproduction is something magical and mysterious that happens when a man and woman copulate. Today, and certainly more in the future, cells can be coached and manipulated into embryonic development, or at least, stem cell development, in a variety of ways. In vitro fertilization may be the most common and practiced form of this. Mostly driven by the desire to locate a source of stem cells, however, science labs around the world are crossing new thresholds in creating a kind of artificial conception. There are, for example, efforts to create "parthenogenetic" eggs:

> Embryonic Stem Cells can now be derived from a human parthenote, an unfertilized egg that is stimulated, by chemical or electrical impulse, to start growing like an embryo. It, too, can be grown to roughly blastocyst stage, its stem cells then harvested. Certain species of insects, fish, and lizards reproduce parthenogentetically, with an egg launching into development without a sperm. But it's only lately that scientists have gotten mammalian eggs to take this track. A parthenote allegedly cannot advance to fetal stage, since it lacks paternal DNA that promotes the growth of the umbilical cord. Some scientists reason that because a parthenote is not a product of egg and sperm, it can't be considered an embryo, making it a noncontroversial source of stem cells. Whether a parthenote's stem cells function normally still needs validation.[158]

In other cases, key human components for inception are replaced by those of animals:

> From China comes news of yet another way to derive embryonic cells. A former NIH researcher reports taking human skin cells, transferring their nuclei into rabbit eggs,

and through this inventive instance of cloning, growing human embryos to five day blastocyst stage to retrieve their stem cells. A major advantage of this approach is that it would avoid using human eggs.[159]

In addition, some researchers are developing methods to take multipotent stem cells (adult stem cells) and through chemical interactions, lead them to revert back to a pluripotent state. This process has some inherent drawbacks, however. In her book The Proteus Effect, Ann Parsons states that "even if a mature cell could be demoted to immature and made pluripotent, it would contain mutations that typically accumulate in the DNA of older cells, making it and its progeny less than ideal for cell therapy"[160]

As can be seen by these examples, embryonic development can be accomplished through a variety of techniques. If one believes that life begins at the moment of conception, is this notion attached to a natural form of conception that no longer holds full court? One certainly could oppose these other forms of embryonic development, as does the Catholic Church. In the June, 2006 document *Family and Human Procreation*, the Catholic Church re-affirmed its position that "the human being has the right to be generated, not produced, to come to life not in virtue of an artificial process but of a human act in the full sense of the term: the union between a man and a woman."[161] Regardless of whether the Church opposes it or not, "produced" embryos (some of which have no potential for total development) are becoming reality. All of us, including the Catholic Church, will have to wrestle with the life status of these engineered embryos. As we do, the whole question of the status of life for the fertilized egg must be debated.

To some degree, our final measurement of stem cell research to the compass of God's stipulated freedom has a "wild card" factor much like we witnessed in the consideration of the bioengineering of animals. Without resolution to the question of the life-status of the human embryo, the arrow of the compass could conceivably swing in very different directions. If an eight day old embryo is a human life, then that life is deserving of God's values of love and justice. If it is merely a piece of nature that has the potential of becoming a life, much like any other cell put under the mechanics of cloning, then it is part of the nature that God has given us permission to manipulate.

In the end, it seems that all of this debate will pass. On the one hand, researchers are continually searching for good sources of pluripotent stem cells that do not require the destruction of an embryo. Just a month before this writing, researchers at Wake Forest University announced the discovery of stem cells in amniotic fluid.[162] Although these stem cells look promising, it is generally believed that they are not as pluripotent as embryonic stem cells. Even so, they may be far more versatile than adult stem cells, and thus provide a good alternative to embryonic stem cells. Time will tell.

Meanwhile, the Bush Administration has allowed Federal funding for stem cell research on a limited supply of stem cell lines. It is very conceivable that in the future, life saving therapies derived from this research will give patients around the world hope and healing. In some cases, stem cell therapies will become the conventional way of treatment. Needing that treatment could be someone who is currently wrestling with this whole issue, and quite unsure about whether embryonic stem cell research was morally right. In that hospital bed of the future, the question will not be so much "is stem cell research ethical", but rather, "Are stem cell therapies *morally tainted*?"

There are some today who are already planting those questions in the minds of people who could possibly be in that hospital bed of the future. In responding to President Bush's announcement of limited funding for stem cell research, Ben Mitchell from the Ethics and Liberty Council of the Southern Baptist Convention wrote:

> Make no mistake about it, these cells have been harvested by killing human embryos. They are morally tainted and any benefits from research on those cells will be ill-gotten gain...Our tax dollars should not be used to fund research we find morally reprehensible. Yet, President Bush's decision makes us pay for tainted research. It's like forcing us to eat our own offspring and charging us for the meal.[163]

If these words remain in the minds of good hearted Southern Baptists, they may very well wonder if they should decline life saving stem cell therapies in the future. Even though those therapies may be far removed from the current debates, these people may fear that the therapies are, as Mitchell said, "morally tainted".

What we have seen in this study, however, is that matter (i.e. nature) does not carry value intrinsically. If nature is not intrinsically sacred, then it is also not intrinsically damned. Our theology does not support the notion of morally tainted matter. In an earlier chapter, we briefly touched on this as we considered the case of the Apostle Paul and his dealing with the Corinthian Christians. These Christians were wrestling with a difficult moral dilemma. In converting to the Christian faith, they had abandoned the worship of the many idols that dotted the landscape of ancient Corinth. They knew that the worship of idols was morally forbidden. They also believed that anything derived from the worship of idols would be morally tainted, and therefore untouchable. One of the things that were derived from the worship of idols was meat. Animals were regularly sacrificed at these idols, and their meat entered the food chain of Corinth. This meat, in the view of the Corinthian Christians, was morally tainted. It should be avoided. Yet, this morally tainted meat was a common benefit throughout the city. It ended up in the meat markets, and at social gatherings. A Corinthian Christian would often not be able to know if the meat being offered comes from the idols or not. This was the ethical dilemma that they faced, and apparently they had posed a question to the Apostle Paul that could have come from our imaginary patient in a future hospital room: What should we do with the benefits derived from the morally tainted?

It is fascinating to watch the Apostle work his way through this difficult question. Space does not allow us to offer a full exegesis of his words. He spends a great deal of time addressing the nature of idols, and expanding on his conviction that Christians should be mindful of the example they set for others. But when he comes to the heart of the question, he says:

> Eat whatever is sold in the meat market, asking no questions for conscience' sake; for 'the earth is the Lord's, and all of its fullness.' If any of those who do not believe invites you to dinner, and youdesire to go, eat whatever is set before you, asking no question for conscious' sake.[164]

Here we find a Biblical model for addressing the benefits derived from the morally tainted. According to this ethic, the moral stain is not so much carried in the actual meat, but rather in the consciousness of the individual. Paul recognizes that this benefit (i.e. meat) was derived from the worship of idols, but once it becomes the meat of the food market, it is no longer branded with that relationship. In other words, morality isn't a kind of virus that carries forward in forms of nature. It isn't akin to Mad Cow Disease. The derivative of idol worship (i.e. meat) does not carry an ethical infection. In fact, that meat is a part of the fullness of God's fertile earth. And the earth belongs to no idol. "The earth is the Lord's."

In the future, the stem cell controversy will subside. Most likely, we will never solve the major controversy to everyone's satisfaction. Questions surrounding the ethics of the pioneering days of stem cell research may linger. Stem cell therapies, however, will pour out remedies of love and justice around the world. Those remedies will not be morally tainted. They will be agents of God's grace, a part of *the world as sacrament.*

For those who believe that the destruction of embryos is murder, I will leave it to you to design the compass. Most likely, the arrow will be leaning south. In this book, notwithstanding the unresolved questions, the compass will remain as it is in figure 15.[165]

We have concluded our analysis of stem cell research in light of the compass, but not all embryonic manipulations are for the purpose of harvesting stem cells. As we move into the future, more and more prenatal engineering will occur with the primary purpose of designing the life that will eventually unfold. At this point, it is time to shift toward the second section of this chapter. It is time to move from Hermes to Hercules.

Prenatal Engineering

In Greek mythology, Hercules was a stunning figure. One of the most popular figures of the pantheon, Hercules was known as an ultimate athlete and warrior, and was credited for beginning the Olympic Games. What gave Hercules such extraordinary abilities? His genetic makeup held the clues. He was mortal due to the genes of his mother. Yet his DNA was comprised not just of mortal genes, but god-genes as well. His father was Zeus.

Hercules was a rare case, and not many others could claim such special genes. What if, however, Zeus genes were available as a commodity, and procedures existed in which those genes could be engineered within the genome of any embryo? If this were the case, there may have been many more Hercules running around. Such possibilities would have been nothing more than the fodder of myth in earlier times. We are entering an era, however, in which this possibility is looking more and more feasible. The subject of prenatal engineering is the next issue to which we will apply the compass.

History will be the final judge, but there is a possibility that a discovery made in our own day will be considered one of the most important of all time: the sequencing of the human genome. That, however, was just the beginning of the journey. The sequence text of the genome was three billion letters long, written in what the director of the project, Francis Collins, called "a strange and cryptographic four letter code"[166] Ever since the genome was rolled out for public view in the year 2000, researchers have been working to decipher that code and identify the role and function of each gene. As the secrets of the genes are unlocked, researchers are finding root genetic causes for various illness and diseases, as well as many human traits and characteristics. The work is tedious and never-ending. In one example, Francis Collins describes his investigation into the genetic causes of a condition known as "hereditary persistence of fetal hemoglobin", in which a normal and necessary transition of blood oxygenation from the fetus stage to the post-birth stage in which babies utilize their own lungs fails to occur. Finding a genetic cause for this problem could lead toward therapies to treat sickle cell anemia. For this one, isolated attempt at locating something on this huge genetic map, Collins writes:

> I will never forget the day when my sequencing efforts revealed a G instead of a C in a specific position just 'upstream' of one of the genes that triggered fetal hemoglobin production. This single letter alteration turned out to be responsible for leaving the fetal program switched on in adults. I was thrilled but exhausted-it had taken eighteen months to discover this single altered letter of the human DNA code.[167]

This testimony of Francis Collins reveals two things. First, the key to understanding, and possibly controlling, human traits, characteristics, malfunctions, and diseases lies in the human genome map that has been discovered. The second lesson, however, is the reality that it will take a long time, with huge investments of time and money, before this map is deciphered.

The work, however, is ongoing. Promising discoveries are being announced all the time. Genetic links have been discovered in Down syndrome, cystic fibrosis, dwarfism, breast cancer, fragile X syndrome, Huntington's Disease, Duchenne muscular dystrophy, and various types of nervous system degeneration.[168] Already, expectant parents can screen their unborn babies for these diseases. In this early stage of genetics, screening is all that is possible for many conditions. The biotechnological goal, however, is to do far more than observe genetic realities. The goal will be realized when we have the ability to splice unwanted elements out and insert desired elements in to determine a different outcome from that which nature has started. This is the essence of prenatal engineering. By and far, it is still an unrealized dream of the future. Genetics as a field, however, is making tremendous strides. Before we know it, prenatal engineering, in at least a limited fashion, will be actualized. Does prenatal engineering fit within God's stipulated freedom? Many people who oppose the notion of prenatal engineering do so because they believe that we have no right to change the course of nature's determination for a fetus. Doing so would be "playing God". Once again, we must assert the basic freedom upon which our compass is built. The actual manipulations of DNA to effect a change in a developing human fetus is, in and of itself, within God's stipulated freedom. DNA is not sacred ground. There are things, however, that are sacred in prenatal engineering. They are the core values of God: love and justice. Let us now hold the compass to the road of prenatal engineering, and see which way the arrow turns.

Movement toward Love and Justice

As we move toward the evaluation of this subject according to the compass, it is necessary to differentiate between two related, yet separate issues: *prenatal screening* and *prenatal engineering.* Prenatal screening is the ability to observe genetic characteristics, yet perhaps without the ability to change them in any way. This would allow perspective parents, and  others, to know the status of an unborn child in terms of disease potential and other information. Screening alone does raise a number of ethical questions related to the core values of love and justice. Should a baby be aborted because of a less than optimum genetic screening? Should insurance companies and prospective employers know screening results? The list of questions can grow, and are being debated in many forums. On May 1, 2003, a comprehensive bill was introduced for legislation in the United States Congress. Known as H.R.1910, this bill sought to prohibit genetic discrimination, particularly by insurance companies. The latest status report indicates that the bill is tied up in bureaucratic processes.[169]

Although prenatal screening is an important subject to address, the subject chosen here goes an important step beyond. It is the hoped-for ability to not only observe genetic characteristics, but to actually change them. If this biotechnological dream is realized, how would such genetic engineering measure up to the core values of love and justice?

The first observation is that the magnetic pull toward love and justice is, indeed, very strong. If biotechnology offers us the opportunity to discover conditions that will impede a totally healthy development of a life and to change those conditions, God's justice is hampioned. In this act, the resources available work to establish an equality, an even playing field, for human life. No longer would one child have healthy blood and another sickle cell anemia; no longer would one child have a healthy nervous system and another spina bifida; no longer would one child have a normal chance of learning and another an inhibited chance. Prenatal engineering could be a wonderful tool to enact equality in the overall health potential of human progeny.

We begin, then, claiming the great hope and promise of prenatal engineering. As a pure, unadulterated concept, it aligns very well with God's stipulated freedom. If we could leave it there, that would be nice. Unfortunately, this issue does get some significant pull from other directions.

Movement toward Significant Risk

With the initial revelation of the human genome, there was tremendous excitement about the potential of mastering it. It seemed quite simple: find the gene that causes a condition. If this gene is needed

but missing, insert it. If it is present and causing damage, remove it. We were on our way to engineering a human life in the same way an architect would design a building. Early on in this not-so-long history, some people wrote books raising these high expectations. Matt Ridley, a British science writer, released his book Genome: _The Autobiography of a Species in 23 Chapters_, in 1999. The title of the book illustrates the notion that the human genome is quite easy to get our arms around.

As time went on, however, the hopes for rapid genetic engineering abilities have become muted. It is becoming clear to researchers that the human genome is an incredibly complex structure, and that the idea of "one gene, one feature" is rapidly dissipating. For many characteristics and conditions, genes alone do not tell the whole story. Environmental factors can often play significant roles.

It is also becoming clear that the more we know, the more we discover that we don't know. Genes are proving to have complex interrelationships, some of which might not be discovered until after an engineering intervention has taken place. A good example of this possibility is something that has been discovered concerning an illness that we have already discussed sickle cell anemia. On the surface, it may look simple and desirable to find the genetic cause of this disease and remove it from the human genome. It has now been determined, however, that the same gene, balanced with a gene of healthy hemoglobin, provides a protection from a disease that causes an even greater risk to the total population, malaria. The Harvard University website sums up the situation this way:

> People with normal hemoglobin are susceptible to death from malaria. People with sickle cell disease are susceptible to death from the complications of sickle cell disease. People with sickle cell trait, who have one gene for hemoglobin A (normal) and one gene for hemoglobin S (Sickle Cell), have a greater chance of surviving malaria and do not suffer adverse consequences from the hemoglobin S gene. [170]

Who would have guessed this relationship prior to the genetic studies? How many more relationships exist, some even more elusive to our own sense of rationality, which remain hidden in the secrets of our genes?

Given this, there is a clear element of significant risk in the field of genetic engineering. This risk is magnified by the fact that genetic alterations can become a permanent feature of an ongoing genetic line. If an irreversible error is made, it could affect generations down the line. In genetic engineering, we are assuming that we can direct evolution better than nature can. In some cases, we most likely will be proven right. There is a risk, however, that we may make mistakes, with the potential of large scale ramifications.

This risk, along with early animal experiments that have shown that things can quickly go awry, led the Professor Emeritus of Biology at Harvard University, Dr. Ruth Hubbard, to publicly oppose prenatal engineering. She wrote:

> Objections center on the fact that a genetic alteration introduced into an embryo is likely to become a permanent part of the genetic endowment of the person into whom that embryo develops and thus also of all of her or his progeny. Considering that the procedures themselves are experimental and the results are unpredictable (laboratory mice on which such procedures are performed often produce progeny with malformations, behavioral abnormalities, or increased cancer rates), germ-line genetic engineering poses unacceptable risks for "persons" who have just barely been conceived. There is no justification for undertaking such manipulations.[171]

It is clear how Dr. Hubbard would point the arrow of the compass. Even if we hesitate to join her statement that there is "no justification" for prenatal engineering, we must recognize that significant risks are present, and the pull toward the west of our compass would be significantly strong. There may be, however, an even stronger pull past west, and toward the south. It is time to consider the opposite polarity to God's core values of love and justice.

Movement toward Increased Imbalance or toward Selfishness

Life is an uneven playing field. There is a standard of desirability that we might label "normal", or perhaps, "healthy".

Whenever we fall short of that standard, we are, in some way, less than whole. We stand in need of correction. We stand in need of *therapy*.

Although we have noted some significant risks to the enterprise of prenatal engineering, we began by affirming the promising potential that genetic engineering offers to restore wholeness, and therefore equality, to those who need such therapy. In this way, prenatal engineering could be a great tool of justice in the world. The same technologies, however, could be employed to go beyond this action. Prenatal engineering could, in fact, begin with an embryo that is perfectly healthy and normal, and proceed to endow that embryo with qualities or characteristics that go beyond: beyond health, beyond normality, beyond therapy.

On November 28, 2001, President George Bush commissioned a team known as the President's Council on Bioethics. The Council studied the various issues of biotechnology and ethics for nearly two years, and finally published their report in 2003. In articulating the fault line upon which ethics shift in the total arena of biotechnology, the council settled on a title to their report: Beyond Therapy . The report states:

> Though we shall ourselves go beyond this distinction, it provides a useful starting place from which to enter the discussion of activities that aim 'beyond therapy.' 'Therapy', on this view as in common understanding, is the use of biotechnical power to treat individuals with known diseases, disabilities, or impairments, in an attempt to restore them to a normal state of health and fitness. 'Enhancement', by contrast, is the directed use of biotechnical power to alter, by direct intervention, not disease processes but the 'normal' workings of the human body and psyche, to augment or improve their native capacities and performances. Those who introduced this distinction hoped by this means to distinguish between the acceptable and the dubious or unacceptable uses of biomedical technology: therapy is always fine, enhancement is, at least prima facie, ethically suspect.[172]

The report goes on to state that the black and white distinction between therapy and enhancement in terms of ethical acceptability is not a perfect measurement. In many ways, it states, enhancement to nature has been a hallmark of the progression of humankind. In this study, we too would not draw an automatic line between therapy and enhancement. We begin in the realm of freedom. If human engineering of a genome could increase physical or cognitive strength beyond nature's prescription, that could be a good thing in which God would delight. The President's Council reported that "over the past few decades, researchers have identified *single* gene alterations that, in a number of species, dramatically extend life."[173] The combination of these and other processes of age retardation have resulted in increased life spans of up to 75%. If these experiments lead to human life spans that exceed 150 years, this could be within God's stipulated freedom. There is no line to be drawn between therapy and enhancement in terms of the actual manipulations of nature.

The real question, however, is how these manipulations would align with the stipulations that they be exercised in the context of love and justice. It seems almost a given that there is one entity that would make a clear distinction between therapy and enhancement: insurance companies. Without insurance coverage, private or public, these enhancements would not be possible for those who do not have the economic means to pay for them. These treatments, then, would become a commodity rather than a therapeutic intervention. A small fraction of the world's population would benefit. As noted, there is already a large imbalance of equality in this world between the rich and the poor. If prenatal engineering becomes an enhancement tool for the wealthy, this imbalance would be greatly increased. We would create a class of super beings; people with super-physiques, super-brains, super-lifespans, etc. Unless this technology was somehow offered to the general population, it would be antithetical to God's core value of justice. The very nature of designing "super" children, however, would cause those who have this ability to resist any effort to spread the blessing. One could not be "super" unless that superiority is measured against a lower standard. The very concept of designing children, beyond therapy, is to give them an advantage over others. This is, at the core, selfish behavior. Prenatal engineering as an enhancement for the rich is antithetical to the core value of love as well as the core value of justice. The arrow would be pulled strongly to the south.

How then, should we ultimately envision the subject of prenatal engineering against the backdrop of the compass? Perhaps once again we have to draw the compass in two different ways. *If* prenatal engineering serves the primary purpose of therapy, and *if* any use of prenatal engineering for the purpose of enhancement is driven with a passion to make the enhancement widely accessible, then the remaining problems reside in the presence of significant risks. The compass could be drawn as follows (fig. 8):

Love & Justice

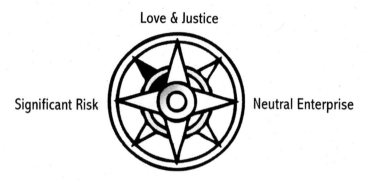

Significant Risk   **Neutral Enterprise**

Increased Imbalance Selfishness

Figure 8

If, however, prenatal engineering becomes the commodity of the wealthy to create progeny with hugely significant advantages over the general population, leading to generational differences as these divergent genomes are passed on, the assault to God's core values of love and justice would be severe. The compass (fig. 9) would reflect that:

Love & Justice

Significant Risk   **Neutral Enterprise**

Increased Imbalance Selfishness

Figure 9

In the emerging abilities of genetic engineering, the human community has some significant choices to make. How we establish guidelines and protocols for the use of this technology will largely determine whether we move forward with God's blessing and within God's stipulated freedom, or whether we slip deeper into an estranged relationship with our Creator.

Chapter Conclusions

In this chapter, we have applied the compass to two major areas of human bioengineering. In each case, the delineation between the non-sacredness of nature and the sacredness of love and justice has been affirmed. On the one hand, viewing biotechnology in the light of God's gift of freedom is a wonderfully liberating concept. It is, perhaps, a place where religion and science could meet. There is no need to fight over matters of biology. Let freedom ring!

On the other hand, we have met our stipulations. To exercise biotechnology within the parameters of love and justice will take a concerted effort that could only be achieved by several entities working together. Religion can articulate the core values; government can develop rules and regulations to enact those core values; science can proceed to confidentially engage its work, knowing that the applications have some commonly accepted parameters. It is interesting to see the convergence of these forces upon something so small: the human embryo. With its stem cells and its genome, the human embryo has become a major stage for one of the most important debates of our time.

Thus far, we have considered biotechnology in its abilities to engineer the carbon based matter that comprises the human being as such is currently known. In addition to this, there is a whole other world emerging that promises to interface with the human form in astonishing ways. With the compass in hand, it is time for us to consider the digital and mechanical revolution that is surely coming to the human body. In the following chapter, the Greek gods *Hephaestus* and *Hades* will lead the way.

Chapter Five Notes

[153] As a resource for this background material, I am indebted to the book The Proteus Effect: *Stem Cells and their Promise for Medicine*, by Ann B. Parsons

[154] icsouthlondon.icnetwork.co.uk/0100news/0200southlondonheadlines/ tm_method=full%26objectid=15179078%26siteid=50100

[155] I wish to thank Dr. Richard Burt, a stem cell research Doctor at Northwestern University Hospital in Chicago, for sharing this view with me as we conversed about this subject.

[156] http://www.health24.com/medical/Condition_centres/777-792-1987-1999,26807.asp

[157] Ibid

[158] Parsons, Ann B. The Proteus Effect: *Stem Cells and their Promise for Medicine.*, p. 241

[159] Ibid, p. 242

[160] Ibid

[161] http://www.siecus.org/policy/PUpdates/pdate0257.html

[162] http://www.msnbc.msn.com/id/16514457/

[163] http://erlc.com/erlc/topics/C24/articles

[164] I Corinthians 10:25-26

[165] See page 142

[166] Collins, Francis S. The Language of God. *A Scientist Presents Evidence for Belief.*, p. 1

[167] Ibid, p. 110

[168] This list comes from an article in Time Magazine entitled "Good Eggs, Bad Eggs.", by Frederic Golden. Time 153.1 Jan. (1999): 51

[169] The status of this bill, as well as others related to the ethics of genetic screening, can be tracked at the Library of Congress website, thomas.loc.gov/cgi-bin/bdquery.

[170] http://sickle.bwh.harvard.edu/malaria_sickle.html

[171] http://www.zmag.org/zmag/zarticle.cfm?Url=/articles/march02hubbard-newman.htm

[172] Kass, Leon (Ed). Beyond Therapy. Biotechnology and the Pursuit of Happiness. *A Report by the President's Council on Bioethics.*, p. 13-14

[173] Ibid, p. 175

6

Enter Hephaestus and Hades

The Compass Applied to Bionics, Artificial Intelligence, and Cybernetic Immorality

As we now cross the bridge from technology that deals with our carbon based bodies to the technology that presents digital and mechanical issues, we meet our next character from Greek mythology, Hephaestus. Born to Hera and Zeus in a way that foreshadowed our new possible ways of creating living stem cells (Hera apparently caused herself to generate a living embryo without Zeus' help!), Hephaestus became the great tool maker of the gods. His tools, however, went beyond any traditional definition. He was able to form and fashion tools and resources that became integrally united with the recipients. It is hard to imagine Zeus without his thunderbolt, Cupid without his arrows, Helios without his magic chariot, or Achilles without his invincible armor. Yet all of these defining elements of the gods were not endowed by nature, they were created by a peer: Hephaestus.

Issue One: Bionics

In the emerging world of biotechnology, we seem to be on the threshold of creating such integrally connected, yet artificial to nature, components. There are promising new developments in the realm of prostheses, which we can identify under the title *bionics*.

Prostheses to replace human parts that have been lost or damaged are, of course, nothing new. Until now, however, these have been very rudimentary and obvious. It is very clear in the story of Peter Pan, that Captain Hook has a fake arm. His hook in no way resembles, or functions like, a real arm. One can easily tell the difference between a real leg and a wooden peg. Sammy Davis Junior could not conceal the fact that he had a glass eye. It filled the socket, but it was lifeless and unmoving. In all these cases, the artificial parts simply filled the space of the real ones. Anyone who looked at people who had them knew that they were, in some fashion, not "whole". Likewise, those who had these prostheses could hardly envision them as part of themselves. The artificial limb or eye was an appendage to the person, and could often be removed and replaced at will.

The emerging world of bionics, however, is promising to revolutionize the realm of prostheses in profound ways. Artificial parts will, more and more, resemble natural ones in remarkable ways. The digital-driven mechanics will allow movement and function akin to the natural. Synthetic coverings for them are beginning to look and feel as natural as real human skin. In the era of bionics, one will not be able to tell the difference between a real leg and an artificial one. This will not only be true about appearance, but also about function. The most amazing development currently underway is the work that is connecting the nerve impulses of the brain to the digital commands of the artificial components. In early experiments, such neural-digital connections are offering promising results.

In 2003, researchers at Duke University successfully connected the brain commands of chimps to mechanical arms. In reporting this story, the *Washington Post* stated:

> Scientists in North Carolina have built a brain implant that lets monkeys control a robotic arm with their thoughts, marking the first time that mental intentions have been harnessed to move a mechanical object. The technology could someday allow people to operate machines or tools with their thoughts as naturally as others today do with their own hands. 'It's a major advance', University of Washington neuroscientist Eberhard E. Fetz said of the monkey studies.

'This bodes well for the success of brain-machine interfaces.'[174]

It seems like a tremendous step forward if artificial limbs can be enhanced through this technology, but it would be even more revolutionary if bionics would allow functioning artificial organs. Some of the earliest work in this field has been in the development of the artificial heart, but current research is taking this issue giant steps forward. Today, many people are benefiting from artificial hearing, through the cochlear ear implant. Even something as complex as the human eye is now being artificially created. Early research in the "bionic eye" has made some significant inroads. Not only will future artificial eyes actually function for vision, but they will move in concert with a natural eye, responding to the same brain command. At the Chinese University of Hong Kong, researchers explain some of their recent efforts at making this possible:

> In this research project, we try to develop a robotic prosthetic eye for people wearing artificial eye implants, which can move like a real eye. After two generations of the prosthetic eye model, here is the latest of the eye implant as shown in the above picture (*sic, picture on the web page*). It detects the natural eye movement of the good eye of the patient who needs to wear a prosthetic eye, using the patient's EOG (electro-ocular-graph) signal taken through a pair of electrodes placed on both sides of the head at the height of the eyes. The built-in control system will then control the movement of the robotic prosthetic eye to follow the movement of the good eye of the patient with a tracking error smaller enough so that people will not be able to tell the difference of the movements of the real and prosthetic eyes. This project shows a good example of applying modern electronic engineering in medical applications to help people in need.[175]

If Sammy Davis Junior lived in the second half of the 21st century, his artificial eye might have been his own secret.

Currently, medical science produces basic artificial components with mechanical functioning, such as the artificial hip or knee. Emerging on the scene now, are artificial replacements for sensory functions such as hearing and seeing. In the not too distant future, fully functioning artificial limbs will be the secret of their owners. All of these functions, however, are driven by the most basic organ in our bodies: the human brain. The brain is the seat of our physical functions, our emotions and memory, our self-awareness, and perhaps even our spirituality. We may be able to see ourselves as human beings augmented by artificial limbs and sensory organs. We could even add internal organs like hearts and livers to that list. The brain, however, seems to us to be a kind of "holy ground" of the human person. Substituting our biological brains with digital and mechanical replacements would, for many people, cross a threshold of what it means to be human.

It is beginning to seem, however, that the brain is, itself, a candidate for bionic replacement. In 2003, news was released to the public concerning experiments with an artificial hippocampus. This part of the brain controls the important functions of memory and learning. The device is like a computer chip, bridging nerve impulses going to the hippocampus with other parts of the brain through electrical currents. The March 12, 2003 edition of the *New Scientist Magazine* reported the story:

> Imagine that one could replace part of a damaged brain with a computer chip that perfectly replicates the brain. Imagine that one could use instrumentation to "read" brain activity, and use that information to program a computer chip. Imagine, then, the downloadable brain. The world's first brain prosthesis - an artificial hippocampus - is about to be tested in California. Unlike devices like cochlear implants, which merely stimulate brain activity, this silicon chip implant will perform the same processes as the damaged part of the brain it is replacing. The pending experiments will test whether the computer chip will indeed function effectively as an artificial hippocampus. Eventually, Berger believes, laser based chips that replicate 10,000 neurons in an implantable component the size of a peanut will be possible. If so, imagine what may be possible when nano-scale, quantum computing becomes a reality, if it does. Work is also being done on developing processes by which the dendrites and axons of surrounding brain tissue (the brain's communication conduits) simply grow themselves onto the artificial chip.[176]

We do not yet know where this research will eventually lead, but it opens the door for possible implications that could revolutionize life beyond anything ever before experienced. Some of these possible implications will be discussed later in this chapter, when we visit with the god Hades. For now, it is lifted up as an example of where the emerging field of bionics can take us. Now that we have a picture of the bionic possibility, it is time to put it to the test of our compass. Is this within God's stipulated freedom?

Movement toward Love and Justice

It is not hard to imagine the abundant ways in which bionics fulfill God's core values of love and justice. There is, of course, the question of access. Every aspect of biotechnology will have to wrestle with the imperative to make the benefits available for the masses. Given that, however, bionics as an advancement in our abilities  to replace damaged or missing parts enacts love and justice in admirable ways.

It is important to note, however, that as bionics offer more and more alternatives to biological parts, Christian theology will be challenged. Stated in a more specific way, Christian theological *anthropology* will be challenged. As stated in chapter one, our theology has seen humanity as a special creation, formed in the image of God, placed in nature but qualitatively separated from nature. Our sense of humanity's identity is integrally connected to our biology. One could assume, certainly, that our theology could easily absorb artificial parts that augment a human body. If, however, bionics ultimately takes us to a place in which the brain is artificial, the loci of our sense of existence will be replaced. From there, it is not too hard to conceive of a "brain wave" being freed from the biological body all together. If neural brain functions can be digitized by an artificial hippocampus, it seems possible that they could be bridged to external devices as well as other parts of the brain or body. This may seem the stuff of science fiction to some, but the technology is not that far fetched. In a bold effort to demonstrate this, and seen as an irresponsible publicity stunt by others, a researcher used his own body as a guinea pig. Dr. Kevin Warwick, professor of Cybernetics at the University of Reading, England, decided to create a neural-digital interface between his own brain functions and external components. Although it is a long quote, let us view an article about Dr. Warwick's experiment as it was recorded in the March 13, 2003 *Dallas Morning News*. The quote is offered as an illustration of the possibilities of human intelligence *leaving* the human body, a point central to the rest of this chapter:

> For Kevin Warwick, becoming a cyborg is the only logical way to deal with failures of the flesh. Unlike computers, his brain can't multi-task, he reasons. Unlike hospital X-ray machines, he can't see through skin. And, unlike electronic communications, his voice seems slow and inefficient to him. 'Look at all the wonderful things machines can do that we can't do,' the professor of cybernetics told a rapt audience at the SXSW Interactive Festival this week. 'And why not? We have the technology. My attitude is, 'Well, let's have a try.' And so, Dr. Warwick last year arranged for a tiny array of 100 electrodes to be surgically implanted into the largest nerve running down the inside of his left arm. For three months, he and a team of 20 scientists from across the globe connected Dr. Warwick's nervous system to computers, a robotic hand and dozens of other gadgets. In essence, he became the world's first real-life cybernetic organism – part human, part machine. The results were as startling as the experiment itself. Hooking several pins into a wireless transmitter, the 49-year-old professor was able to crudely control the motions of a nearby robotic hand. In the same manner, he could quickly jerk his hand closed to control the motions of an electric wheelchair, turn on a light and navigate through a simple computer desktop interface.
>
> And, in one of the strangest twists, Dr. Warwick's wife, Irena, volunteered to have a less complicated implant placed in her own left arm in an attempt to electronically communicate with her husband. "When my lovely wife moved her hand three times, I felt three pulses of current," he reports. "It was sort of like Morse code between our nervous systems." Some scientists scoffed at Dr. Warwick's guinea pig tactics. Religious zealots condemned the experiments as

violations of the body's sanctity. But Dr. Warwick eagerly
argues that we humans are already well on our way to
becoming cyborgs. Artificial hips, knees and other joints are
in widespread use today, he points out. And doctors now
routinely repair hearing difficulties using cochlear implants.

To Dr. Warwick, the next logical step is to route around
the nerve damage that causes paralysis. If nervous system
messages can be wired for transmission to other cyborgs or
devices, suddenly a whole new world would open for the
disabled, he says. "You're looking at driving an automobile;
you could drive and navigate just by thinking about it," he
says. "Somebody who is paralyzed could make coffee just by
thinking about it." Education could be streamlined and
simplified if a professor's knowledge could be imparted to
students by transmitting a bundle of thought waves, he says.
'Why can't I just go, thwack! There you go; you've got it?' he
asks. Speaking different languages could become as simple as
turning a switch mounted in your skull. 'And,' he says, 'I
would love to have my brain linked into the network. I would
love to browse the Web not by pushing buttons, but by
thinking about it.' Cyborgs will have incredible advantages in
society, he says. 'I want to be there amongst the cyborgs,' he
says. To that end, Dr. Warwick says he is seriously
considering having cranial implants that can advance research
into thought and memory transmissions. 'As far as being a
cyborg in the future,' he says, 'I'll be back.'[177]

How would our Christian anthropology understand the presence
of human thought expressed not in the biological body nature has
provided, but in digital impulses and mechanical operations?
Although it is beyond the scope of this book, much more work needs
to be done in our theology of humanity in light of the emerging world
of bionics and "cyborg" possibilities, as well as the related field that
we will consider momentarily, artificial intelligence. Before we leave
the story of the *Dallas Morning News*, it would be a good moment for
us to reassert the bold claims of our study. The report mentioned that
"religious zealots condemned the experiments as violations of the
body's sanctity." In light of this study, we would take exception to
these religious leaders. The human body, as a piece of nature, is not
cred. In God's bold yet stipulated freedom, there is permission to

alter, augment, or even abandon that body. Bionics and cybernetics are to be evaluated not on the merits of the manipulations of nature, but rather on the principles of God's core values of love and justice.

If the field of bionics finds strong magnetic pull toward the core values of God, is that how we would ultimately draw the compass? The slow and careful way in which bionic parts are being tested and implemented do, for the most part, mitigate significant risk. Dr. Warwick may be a good example of someone engaged in a neutral enterprise, although one could quickly point out the possible therapeutic benefits of his work. The major challenge, however, to the overall issue of bionics comes from the opposite polarity of God's core values.

Movement toward Increased Imbalance or toward Selfishness

In the early generations of bionic technologies, the pull toward this polarity will most likely not be very strong. Most people would much prefer keeping their natural and working body parts, and the medical establishment will generally see bionic replacements as a therapeutic remedy to be used only when the original part is damaged or missing.

As time goes on, however, bionics may provide abilities that go beyond nature. Developing a bionic eye that can see and function as well as a human eye is the challenge for today. In the future, however, our artificial eyes may be equipped with the functions of our cameras: powerful zoom lenses; night vision; etc. If artificial elements offer advantages over original ones, will there be a temptation to replace a perfectly good eye with an artificial one? On a much smaller scale, it happens today. Breast implant surgery is a booming business! If such a world of 'consumer bionics" were to emerge, it would most likely be (akin to prenatal engineering) the exclusive realm of the rich. This would contribute to an increased level of inequality in the world. On the one side, the rich would selfishly benefit from the growing powers of digital-neural interfaces, while on the other side the poor struggle in their carbon-based bodies of imperfection.

Once again, the dividing line between justice and injustice is in the demarcation offered by the President's Council on bioethics:

therapy versus enhancement. As mentioned in the last chapter, it is not enhancement in and of itself that falls out of line with God's stipulated freedom. The challenge of biotechnology that enhances is in its likelihood to increase injustice and selfishness in the world.

A selfish use of bionics can be observed in another way that raises a host of question concerning love and justice for both individual persons and the global community. Earlier, we observed the promising new research into the development of the artificial hippocampus. This artificial brain component interfaces digital chips with neural connectors. Theoretically, it could serve as a port to both upload brain activity to an outside source and to download information from an outside source. Multiple brain implants could allow this flow of information to occur among group members, perhaps controlled by a central authority. What could this accomplish if it were employed by, say, a military? An army of "cyborg" soldiers could have instant communication with leadership, in which important commands and battlefield intelligence is digitally transmitted to the soldiers on the ground. If this seems like a far fetched application, it would be sobering to realize the source of funding for the development of the artificial hippocampus: The US Office of Naval Research and Defense Advanced Research Projects Agency.[178] While it is certainly possible that the military interest in this technology is for therapeutic purposes, one can not dismiss the military implications. It may be a slow process for the military to purposely augment its human soldiers in this way, but other creatures seem to be fair game. On March 13, 2006, the *Washington Post* reported that:

> The Pentagon is seeking applications from researchers to help them develop technology that can be implanted into living insects to control their movement and transmit video or other sensory data back to their handlers. In an announcement posted on government Web sites last week, the Defense Advanced Research Projects Agency, or DARPA, says it is seeking 'innovative proposals to develop technology to create insect cyborgs,' by implanting tiny devices into insect bodies while the animals are in their pupal stage.[179]

A military that is showing interest in cyborg ants and human brain implants does, at least, raise some questions.

How would one evaluate this manipulation of nature in relationship to God's core values of love and justice? Like many other forms of technology, bionics would fall into the general net of war and defense. Both before and after Thomas Aquinas developed his "Just War Theory" in the *Summa Theologicae,* debates have raged concerning whether war ever falls in line with God's permission. We can not solve that puzzle here. We can, however, observe something rather unique in the utilization of bionics to augment human soldiers. Equipping a soldier with a gun is one thing. Replacing a portion of that soldier's brain, or removing a soldier's eyes to fit him or her with fantastically capable artificial eyes, would be quite another. In doing this, the purpose of manipulating the nature of that soldier's body is not for his or her sake. The military would be acting for its own good, not for the good of the "other", who is the soldier. This would be antithetical to God's core value of Agape love.

Admittedly, the subject of military cyborgs may seem as science fiction to some. If so, the rest of this chapter may push the envelope of comfort a bit more. It does seem certain, however, that two realities will happen in this 21st century. First, biotechnology will make cyborgs possible. Second, the military could have large strategic benefits in having cyborg soldiers. The rest of the story is yet to be written.

In our overall consideration of bionics, then, we have seen the following: There is tremendous hope and promise in bionic technologies as forms of therapy for many people. There is the strong possibility that bionics could go beyond therapy, and actually serve to enhance humans in new and amazing ways. The arrow has a strong pull north. There is the possibility, however, that bionics will become a commodity of the rich, or the weapon of a military. These possibilities tug at the arrow from the south. As the human community moves forward in claiming these new powers, it is important that the applications of bionic capabilities be governed by ethics that seek to promote love and justice. The temptations will be strong. Yet the good that bionics could bring to the world seem worth the *risk.* In drawing the compass, then, it might be prudent to demonstrate the arrow moving toward God's stipulated freedom, yet experiencing the risk of wrong applications. Perhaps to highlight the presence of the tug to the south, a magnet could be offered: (fig. 10)

Love & Justice

Significant Risk

Neutral Enterprise

Increased Imbalance Selfishness

Figure 10

Issue Two: Artificial Intelligence

Our shift from the subject of bionics to that of artificial intelligence is more of a move across a spectrum rather than an abrupt change of topic. The two are much related. Hephaestus remains the appropriate image for this subject, and does, in fact, demonstrate the progression from making tools that augment people to building true and independent intelligence. As his building capabilities grew, Hephaestus eventually decided to create an entire being. According to the legend, he created the first mortal woman, Pandora. She was, you might say, the first creation of *artificial intelligence.*

During the last decades, much time, effort, and money has been poured into the efforts of creating artificial intelligence. Although there were some initial hopes of early breakthroughs, the quest for artificial intelligence has been a frustrating one. In his book <u>On Intelligence: *How a New Understanding of the Brain will Lead to the Creation of Truly Intelligent Machines,*</u> the computing pioneer Jeff Hawkins explains the problems that have been encountered in the efforts to create intelligent machines. To date, Hawkins states, efforts to build artificial intelligence have been focused on *computation.* The realities have shown that it takes exhaustive commands and programming to enable computers to compute the steps necessary for even a small activity such as catching a ball. New discoveries about the human brain, however, have revealed that most of our cognitive

functioning is not about computation. It is, rather, based on *memory*.
Hawkins writes:

> Let me show, through an example, the difference between
> *computing* a solution to a problem and using *memory* to solve
> the same problem. Consider the task of catching a ball.
> Someone throws a ball to you, you see it traveling toward
> you, and in less than a second you snatch it out of the air.
> This doesn't seem too difficult-until you try to program a
> robot arm to do the same. As many a graduate student has
> found out the hard way, it seems nearly impossible. A
> computer requires millions of steps to solve the numerous
> mathematical equations to catch the ball. A brain solves it in
> a different way. It uses memory.[180]

From shifting from efforts to build artificial intelligence through
computation to building it through memory functions that mirror the
activity of the human brain, Hawkins believes that we might finally
achieve our goal of artificial intelligence. No one yet knows when, or
even if, we will accomplish this goal. Yet if we do succeed in creating
intelligent machines, what issues make that a matter for our compass?
If we do create artificial intelligence, that intelligence could quickly
outpace our own in a number of ways. Artificial intelligence would be
akin to our own brain functions, yet not dependent on our biological
framework. This intelligence would be based on computer processing,
which is far faster than our own. An intelligent machine may not have
a higher IQ than us, but it could evaluate options and choose actions
much faster than we could. In addition, an intelligent machine could
be equipped with the best functions of all the species, whereas we are
limited to the set with which the human species has been equipped.
An intelligent machine, for instance, may have sight and hearing like
us, with the added abilities of sonar and radar. These are the powers
that could be possessed by *one* intelligent machine. Artificial
intelligence, however, would transcend one machine. It could easily
link all such machines together, perhaps linked to a master Artificial
Intelligence that sees all, knows all, and never sleeps. If we are
successful in creating artificial intelligence at this scale, we will have,
for the first time in human history, created an intelligence that
surpasses our own. Let's imagine that this is our destiny. Let us
evaluate this path according to the compass.

Movement toward Love and Justice

The creation of an artificial intelligence could certainly help human beings in virtually every task or enterprise. It could greatly enhance medicine, for instance, by its seemingly limitless ability to research and network. On a personal level, intelligent machines can become servants to humans, much like Rosie, the robot maid of the Jetsons cartoon. We could go on and imagine other helpful functions of artificial intelligence, resulting in establishing a case for drawing the compass with the arrow aligned to the north.

When one stops to consider the total ramifications of an artificial intelligence, however, there seems to more *need* created for love and justice than an actual fulfillment of love and justice. This becomes focused on this important question: In creating intelligent machines, will we create intelligent life? This will become especially poignant if our intelligent machines derive emotions, self-awareness, and artificial bodies that resemble our own. Our new creations will begin to seem less like machines, and more like living things. They will become a new category of "Thou" to our "I". They will be our "other". [181]

As an "other", intelligent machines can become a new class of individuals to whom we are responsible for the proper administration of God's core values of love and justice. In light of this, we would have to develop an understanding of the rights of intelligent machines. We would have to protect them from forms of misuse and abuse. Given the plethora of pornography that has taken over the internet, it seems quite likely that the sex industry would seize intelligent machines for a new era of prostitution. Such "robosexuals", as they are called, could be given such attractive forms that they would be virtually irresistible. The industry would probably insist that these machines are just sophisticated sex toys. A higher ethic would have to be in place and enforced. Henrick Christensen, Chairman of the European Robotics Network at the Swedish Royal Institute of Technology and founder of that institutions "robo-ethics" board, has stated that he believes that the first line of robosexuals will be in use within five years.[182] He is urging the development of a set of ethics now, even though intelligent design is still a thing of the future. In a similar vein, Gianmarco Verruggio, a roboticist at the Institute of Intelligent Systems for Automation in Genoa, northern Italy, has worked with a group of scientists and ethicists to develop a set of ethical guidelines, both for people making robots and a "built in" set of ethics within the robots themselves.[183] The creation of intelligent machines will necessitate a tremendous amount of ethical reflection, and the enforcement of guidelines to protect our new neighbors.

Although our intelligent machines will need protection from us, there are those who fear that we will need protection from them. In our effort to create intelligent machines, we might create something that can assume power over us. On the one hand, we as humans might have to be selfless enough to allow such a transition to happen. In his book The Blind Watchmaker, Richard Dawkins presents a theory of how life could have originated on the planet. He suggests that it is quite possible that RNA was a support mechanism for crystals, but RNA soon became more efficient than the crystals and took over the primary evolutionary role. Dawkins wonders if such an eventual transition will occur between humans and artificial intelligence. Dawkins writes:

> Could it be that one far-off day intelligent computers will speculate about their own lost origins? Will one of them tumble to the heretical truth, that they have sprung from a remote, earlier form of life, rooted in organic, carbon chemistry, rather than the silicon-based electronic principles of their own bodies?[184]

In creating artificial intelligence, might we create a form of intelligence that will, in the survival of the fittest, lead to the end of the era of human domination? If this seems to be the way that life moves forward on this planet, would Agape love require the ultimate selflessness from humanity? Or, perhaps more likely, would we enslave our intelligent machines and limit their own ultimate potential because of our fear of them. Earlier, we noted the work of Gianmarco Verruggio, of the Institute of Intelligent Systems for Automation in Genoa, who has put forward a set of ethics that should be built into every intelligent machine. These ethics include the following: ensuring human control of robots;

preventing illegal use; protecting data acquired by robots; and establishing clear identification and traceability of the machines. Verreggio explains why:

> Scientists must start analyzing these kinds of questions and seeing if laws or regulations are needed to protect the citizen. Robots will develop strong intelligence, and in some ways it will be better than human intelligence. But it will be alien intelligence; I would prefer to give priority to humans.[185]

Similar fears were expressed by Bill Joy, the founder of Sun Microsystems, as reported in the February 16, 2005 edition of *The New York Times*:

> Bill Joy, a co-founder of Sun Microsystems, has worried aloud that 21st-century robotics and nanotechnology may become 'so powerful that they can spawn whole new classes of accidents and abuses.' 'As machines become more intelligent, people will let machines make more of their decisions for them,' Joy wrote recently in Wired magazine. 'Eventually a stage may be reached at which the decisions necessary to keep the system running will be so complex that human beings will be incapable of making them intelligently. At that stage the machines will be in effective control.'[186]

In many ways, more than can be articulated here, the rise of intelligent machines will place new demands for love and justice. These demands will be upon humans in their treatment of the intelligent machines, and perhaps upon intelligent machines in their treatment of humans. All of this would lead us to visualize that the goal of aligning this arrow with God's core values is very difficult. The polarity grows farther away from us as we develop this technology. Likewise, the magnetic pull to the opposite polarity, in which selfishness would reign and inequality would be increased, would be a constant force. Above all, it seems that the strongest pull of all would come from the direction of *significant risk*. In creating intelligent machines, we are introducing new risks into the environment. There are significant risks for the intelligent machines; significant risks for human beings; and significant risks that God's core values would be violated. Given all of the dynamics that are present in the issue of artificial intelligence, the compass must represent a strong pull toward the element of risk. It would, perhaps, look like figure 11:

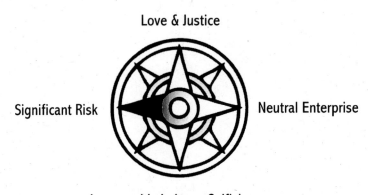

Love & Justice

Significant Risk

Neutral Enterprise

Increased Imbalance Selfishness

Figure 11

The purpose of drawing the compass this way is not to say that efforts to create artificial intelligence should stop. It does indicate, however, that much thought needs to be put into the ethical questions as we progress forward. Failure to do so is, indeed, a very risky enterprise.

Issue Three: Cybernetic Immortality

While some are eager to create artificial intelligence, others believe that the real intelligence that will operate machines in the future will not be artificial, it will be *imported*. As we have already noted, bionics are opening the door to the neural-digital interface. Some theorize that connecting the human brain to a digital platform may allow the important contents of the brain, i.e. memory, emotion, personality, etc, to be digitized. In essence, the "mind" of an individual could be transferred to a digital component, and then uploaded into the external world. Once there, the "mind" is free from the decay of biological demise. In a digital form, this mind could exist forever. It would be a kind of *cybernetic immortality*. Noreen L Herzfeld, in her book In Our Image: *Artificial Intelligence and the Human Spirit,* explains:

Hopes for cybernetic immortality are rooted in the quest for a suitable mechanical platform for the neuronal structure of the brain. Ideas for how the mind might be transferred to such a platform vary. Advances in medical prostheses have led the director of MIT's robotic lab, Rodney Brooks, to predict that human beings and machines will merge in the near future, as we begin to replace more and more of our biological parts, including parts of our brains, with mechanical parts. The ultimate goal is to replace all biological parts, either by uploading a person's neural patterns onto a computer, or by slowly replacing part after part till one becomes all machine. Robotics engineer Moravec, of Carnegie Mellon, describes a similar process of connecting the neurons of the brain to a computer in such a way that, "in time, as your original brain faded away with age, the computer would smoothly assume the lost functions. Ultimately your brain would die, and your mind would find itself entirely in the computer. [187]

Ray Kurzweil, a pioneer of several computer innovations, has made some very bold predictions of how soon and how real cybernetic immortality will be:

There won't be mortality by the end of the twenty first century. Not in the sense that we have known it. Not if you take advantage of the twenty first century's brain-porting technology. Up until now, our mortality was tied to the longevity of our hardware. When the hardware crashed, that was it. For many of our forbearers, the hardware gradually deteriorated before it disintegrated. As we cross the divide to install ourselves into our computation technology, our identity will be based on our evolving mind file. We will be software, not hardware. As software, our mortality will no longer be dependent on the survival of the computing circuitry as we periodically port ourselves to the latest, evermore capable "personal" computer. Our immortality will be a matter of being sufficiently careful in making frequent backups.[188]

Once a mind becomes "software", it can be "loaded" into a variety of forms, from robotic bodies to desk top computers to hand carried devices such as Ipods. This mind could also be linked to other such digital minds, allowing for one to participate in a wider consciousness. As Herzfeld states:

> Through such techniques, the form or organization with which we identify our "I" could be maintained indefinitely, and, which is important, evolve, become even more sophisticated, and explore new, yet unthought of, possibilities. Even if the decay of biological bodies is inevitable, we can study ways of information exchange between bodies and brains which will preserve the essence of self-consciousness, our personal histories, our creative abilities, and, at the same time, make us part of a larger unity embracing, possibly, all humanity; the super organism.[189]

The concept of cybernetic immortality is the final subject to be evaluated by our compass. The Greek god chosen to symbolize this subject is that of Hades, the god of the underworld. The obvious connection of Hades to issues of death and beyond makes Hades a logical choice. There is a deeper reason, however, for his selection. It has to do with the most familiar story about Hades. The Greece visitors' web page sums up that story nicely:

> Hades springs out of the earth and captures Persephone, dragging her off to be his queen in the Underworld. Her mother Demeter searches for her and stops all foods from growing until Persephone is returned. Finally, a deal is worked out where Persephone stays one-third of the year with Hades, one-third of the year serving as a handmaiden to Zeus at Mount Olympus, and one-third with her mother.[190]

How does the story of Hades relate to the subject of cybernetic immortality? Let's put this final issue to the compass, and see which way the arrow swings.

Movement toward Love and Justice

How far would we be willing to push the
case that God has given a fundamental
freedom for us to manipulate nature? If we
totally dispense with the biological body
God gave us, and find a way to cheat what
seems to be a God ordained appointment
with death, would God permit this? Pushing
the assertions of this book to these limits, we

would still offer the now-familiar answer: Cybernetic immortality
would pass through God's stipulated freedom, as long as it promotes
God's core values of love and justice. Does it?

In making a case for cybernetic immortality to be an act of love,
one would have to gravitate to the personal level. It would be difficult
to justify why a certain mind should exist forever, as if the world
could not go on without it. Perhaps we could benefit from an
immortal presence of, say, Einstein's mind. Even Einstein, however,
has proven not to be irreplaceable.

Keeping a mind immortal, then, would be of interest within the
context of love. People who love each other would want never to be
separated. Likewise, people who love themselves might want to
preserve their life, albeit in this very different form. One could say,
then, that it is love that drives the desire for cybernetic immortality.

What kind of love, however, is this? As we observed in an earlier
chapter, the concept of love was expressed in four different words in
the Greek language. The love that drives this desire could certainly be
ἔρως (érōs) love, the love of romance. It could be φιλία (philía) love,
the love of friendship. It could certainly be στοργή (storgē) love, that
which binds parents and children. All of these loves are positive
values, to be sure. The love that is God's core value, however, is
something different: ἀγάπη (agápē) love.

Agape love, as we have learned, is not based on emotional ties.
Agape love is an act of cognition and will. It is marked by a selfless
commitment for the sake of the other. Agape love is stronger, and
more strongly attuned to God's core value, when it is extended to the
one who is not the automatic recipient of our other loves. Agape love
shines when it is extended to the stranger, the outcast, and even the
enemy.

159

In many ways, cybernetic immortality fails to meet the demands of this kind of love. In addition, it is hard to imagine how it would work toward justice in the world. If someone is able to cheat death, how does that contribute to the efforts of equality for the rest of humanity? One could argue, I suppose, that in replacing a carbon-based life with a digital existence, there is one less entity drawing down the limited resources of the world. This might be one way in which cybernetic immortality for the one could benefit the many. Additionally, there is the possibility that this process would become an option for the masses, creating a new space of life somewhere between heaven and earth in which a new burgeoning community is developed. One could even make the case that cybernetic immortality would be a fulfillment of the eschatological hope for humanity, a way of achieving God's promise of an eternal life.[191] If this were to be the case, however, Christian theology would have to embrace some fundamental changes. If the digital world becomes our understanding of the promised heaven, and cybernetic immortality becomes our notion of eternal life, then the realization of these gifts is only afforded to the humans who happen to live late enough in history to benefit from them. In other words, Christian theology would have to surrender the idea of a personal heaven for all believers for an immortality achieved for the human community as a total. Church liturgy and curriculum would certainly have to be rewritten!

One day, we may suppose, cybernetic immortality may be embraced by religious communities, and seen as such a standard choice that no one would perceive it as a particularly selfish thing to do. At this point, however, cybernetic immortality seems to be fiercely individualistic and personal. It is a life seeking to preserve itself. Given these factors, it would be hard to visualize that the arrow of our compass would receive much pull from the direction of God's core values. The pull from the opposite direction, on the other hand, would be strong indeed.

Movement toward Increased Imbalance or toward Selfishness

The desire to be cybernetically immortalized seems, at the core, a selfish desire. It is either a selfish desire of an individual, or the selfish desire of those in relationship with that individual. At this point of view, however, it seems very selfish indeed. Equal to its propensity toward selfishness is its likely contribution to increased inequality. If we are unable to make basic computers accessible to the world's population, what chance is there that something as sophisticated and expensive as cybernetic immortality would be accessible? The human life that is free from biological constraints would have a large array of abilities and possibilities, leaving biological persons far behind. The gulf between the rich and the poor would be widened considerably.

Given all of this, however, it may very well be that the person who is hurt the most by cybernetic immortality is the very one who has entered it. A dislocated human mind is one without a resting place. Even those who celebrate the possibility of cybernetic immortality recognize this fact. Ray Kurzweil states:

> Some philosophers-professional artificial intelligence critics-Hubert Dreyfus for one-maintain that achieving human-level intelligence is impossible without a body. Certainly, if we're going to port a human's mind to a new computation medium, we'd better provide a body. A disembodied mind will quickly get depressed.[192]

Kurzweil believes that providing a mechanical body is the answer to this problem. Like all "software", however, the mind will not be permanently anchored in that body. Kurkweil states that "as software, our mortality will no longer be dependent on the survival of the computing circuitry as we periodically port ourselves to the latest, evermore capable 'personal' computer."[193] A mind that drifts from place to place is still a "disembodied mind", finding at most a temporary residence in some "computation medium". Is this what cybernetic immortality offers: An eternal existence of movement from medium to medium? In the myth of Hades, the woman Persephone has a similar fate. As we saw earlier, Persephone must spend her eternity moving from place to place, dividing her time between Hades' underworld, serving as a handmaid to Zeus, and spending time with her mother. Persephone's world of immortality seems very much like the one offered by cybernetics.

Cybernetic immortality would lead to disembodied minds. A disembodied mind is a restless mind. Ironically, the antidote to this plight is found right where we would expect to find it if we were looking at the compass. The opposite direction of selfishness is Agape love. This love, offered to one's self, could remedy the ill effects of cybernetic immortality. A peaceful eternity, so far as we can know by faith, is found in a *selfless* act. It is the act of giving oneself over to *death*. In death, we are united with God. In death, we are no longer a drifting mind. St. Augustine may have said the best words to evaluate cybernetic immortality:

> *Our hearts are restless until they find their rest in thee, O Lord.*

Time may moderate this opinion, but for cybernetic immortality, the compass is currently drawn as in figure 12:

Love & Justice

Significant Risk   **Neutral Enterprise**

Increased Imbalance Selfishness

Figure 12

Chapter Conclusions

In this final chapter of compass applications, we have considered three major areas of future biotechnology that bring human life into synthesis with machines and digital components. In all three issues, we asserted the God given freedom to manipulate nature. In all three issues, however, significant ethical issues have been identified. As humanity moves forward to realize these dreams, it will be extremely important that the core values of love and justice serve as a measuring stick for each application.

As humanity moves forward into this future, it will also be necessary for Christianity, and other major religions, to do the difficult yet needed work of theological reflection. As noted earlier, one major area of needed theological reflection is on the issue of anthropology, particularly in the *Imago Dei* (image of God). If the future unfolds as imagined in the issues of this chapter, what forms of intelligent life will there be? There will be human minds in biological bodies, human minds in mechanical bodies, artificial minds in mechanical bodies, artificial minds in biological bodies, and a full spectrum of cyborgs with some percentage of biological parts and some percentage of mechanical parts. Are these all created in God's image; If so, *how*? If not, *who*? Failure to accommodate these developments of biotechnology within our theology would lead to that theology becoming as archaic and dead as the Greek mythology that has guided us through these issues.

Developing new ethics based on God's core values of love and justice, while developing new theologies to relate to our unfolding world of biotechnology, become the challenge of the Church in the 21st century. Even if we do that, however, there is still the need to bring those ethics and that theology to bear upon the decisions that will be made. In the concluding chapter of this book, we will consider some ways in which the Christian community could have a positive role in directing the path of biotechnology. Let us now visit the personalities of the Greek Pantheon one more time.

Chapter 6 Notes

[174] http://www.washingtonpost.com/ac2/wp-dyn/A17434-2003Oct12
[175] http://www.ee.cuhk.edu.hk/Admission/eye.html
[176] http://www.newscientist.com/article.ns?id=dn3488
[177] http://www.dallasnews.com/index.html
[178] http://www.eurekalert.org/pub_releases/2003-03/ns-twf031203.php
[179] http://www.washingtontimes.com/national/20060313-120147-9229r.htm
[180] Hawkins, Jeff. On Intelligence: *How a New Understanding of the Brain will Lead to the Creation of Truly Intelligent Machines*, p.68.
[181] Space does not permit a more detailed discussion on the possible ways to imagine the human-machine relationship in the advent of artificial intelligence. For a good overview of these possibilities, the reader is referred to the book In Our Image. *Artificial Intelligence and the Human Spirit*, by Noreen L. Herzfeld.
[182] http://www.economist.com/displaystory.cfm?story_id=7001829
[183] http://www.timesonline.co.uk/tol/newspapers/sunday_times/ireland/article675985.ece
[184] Dawkins, Richard. The Blind Watchmaker. *Why the Evidence of Evolution Reveals a Universe without Design.*, p.158.
[185] As quoted in *Times* Online, http://www.timesonline.co.uk/tol/newspapers/sunday_times/ireland/article675985.ece

[186] As quoted in *The New York Times Online*,
http://www.nytimes.com/2005/02/16/technology/16robots.html
[187] Herzfeld, Noreen L. In Our Image: *Artificial Intelligence and the Human Spirit.*,
p.71
[188] Kurzweil, Ray. The Age of Spiritual Machines. *When Computers Exceed
Human Intelligence* , pp. 128-129
[189] Herzfeld, Noreen L, p. 71
[190] http://gogreece.about.com/cs/mythology/a/mythhades.htm
[191] For more consideration of this view, the reader is directed to the book
Resurrection: Theological and Scientific Assessments, (2002) edited by Ted
Peters, John Russell, and Michael Welker.
[192] Kurzweil, Ray. p. 134
[193] Ibid, p. 129

Conclusion

Becoming Prometheus: Religion Engaging Science in a Positive Way

In Greek mythology, we meet a variety of fascinating characters. Several of these characters have served as symbolic metaphors for the issues of this book. The meaning behind this Greek mythology, however, can become a deeper illustration of the challenges confronting an interface of science and religion in the matters of biotechnology. In concluding our study, the last figures to be invoked are two brothers, Epimetheus and Prometheus. We must once again remind ourselves that metaphors are not perfect. The use of Epimetheus and Prometheus as metaphors is presented here in a limited fashion. The total mythology surrounding Prometheus, for example, would not provide a desirable role model for the Christian community. He is, after all, severely punished with an onslaught of attacking birds. In this chapter, these two figures of Greek mythology are considered in a more narrow scope.

Epimetheus was given a gift from Zeus, a wife by the name of Pandora. Pandora had in her possession a box. It was clear that the box contained amazing powers, but many of them were dangerous to the inhabitants of the world. In this box were the powers that truly belonged to the gods. Pandora was instructed *never* to open the box and then was equipped with an insatiable curiosity. Obviously, having a forbidden box and a cat-like curiosity is not a match made in heaven. Keeping the situation under control was a responsibility that fell to Epimetheus. For a while, Epimetheus was attentive to this task. One day, however, Epimetheus left his house *without thinking* about the consequences of leaving Pandora alone with the forbidden box. As any could predict, curiosity got the better of Pandora. She opened the box, and released all evil upon the earth.

Although that myth is commonly called *Pandora's Box*, much of the myth's meaning is lost if we do not focus on the character Epimetheus. He is the one who had the responsibility of ensuring that Pandora's Box was handled in the right way. Yet he stopped paying attention to that responsibility. He left Pandora, and the box, unattended. Why would he do such a thing?

One of the great things about Greek mythology is that often you don't have to guess about someone's destiny. The clues are contained in their names. So it is with Epimetheus. His name means "after-thought", or perhaps, "hindsight." That was the problem with Epimetheus. He acted before he thought. He took actions before he thought through the consequences of his actions. Because he did, Pandora's Box was opened, and terrible consequences ensued.

The issues addressed in this book have illustrated that humanity is entering an unprecedented era of powers over nature and biology. As we grow in our abilities of biotechnology, we will begin to realize the powers that were, up to now, reserved for "the gods," or more in line with modern theology, *God*. From our ability to master the gene, to the promise of stem cell research; from the advances of artificial intelligence to the realms of nanotechnology; from the development of bionic merging of biology and technology to the leap from the neural to the digital; we are taking our first steps into a brave new world. This "box" of powers that are becoming available to us is truly akin to Pandora's Box, in that it allows us to do things that were once considered off limits to humans. Should this box be opened? If increasing powers of nature are realized, what powers should be actualized, and which ones should remain in the box? Where does one divide the line between what *can* be done and what *should* be done?

These may have been good questions to ask in Epimetheus' house. But the problem was, no one was asking them. Pandora was too overcome by her own curiosity. Just *having* the ability to open the box was all the motivation she needed to open it. In our world today, there are many *Pandoras*. There are, indeed, people in fields of research and development who believe that if something *can* be done, it *should* be done. Many of these scientists live and work in countries with little to no governmental or peer limits. Some of the most rapid and aggressive inroads into what has been considered *sacred space* will come from these sources. In 2002, the Las Vegas based company Clonaid shocked the world by announcing the birth of a cloned human baby outside of the US[194] These claims were never substantiated to the acceptance of the global scientific community, but they continue to be asserted by those who have made the claims. In fact, a year later the same company announced the birth of a second cloned baby, which received similar skepticism.[195]

In addition to the debates about the legitimacy of these claims, there was a large degree of concern over proper ethical boundaries that had been crossed if these claims were true. This indicated that the scientific community does recognize the important responsibility that the "box" of biotech possibilities offers. Yet science is not equipped to ask the important questions of sacred spaces. Science deals with the physical world, not the spiritual world. In other words, science is much better equipped to address the question of *how* a certain road can be traveled. The more fundamental questions of whether or not the road *should* be traveled in the first place are not scientific questions, they are religious questions. The development of biotechnology is a scientific enterprise. The articulation of the compass to guide that development is not. Yet, if science proceeds to open the "Pandora's Box" of biotechnology without thinking through these questions, it has become Epimetheus. When it comes to the possibilities that are on the horizon, with their implications for both good and evil, *hindsight* just won't do. In both Greek mythology and modern biotechnology, Epimetheus is a tragic character.

If Epimetheus is not likely to engage in the forethought that is required before Pandora-like boxes are opened, then who is? In Greek mythology, such a character existed. His name was Prometheus, and he was actually the brother of Epimetheus. His name means "before thought", and like most characters of Greek mythology, he lived up to his name. Indeed, he had warned Epimetheus about the dangers of being a caretaker of Zeus' property. Long before Pandora opened that box, Prometheus had already thought through the possibilities.

In the ongoing dialogue concerning the possible directions and applications of biotechnology, there is an entity that can play the role of Prometheus. It is the entity that has received God's Biblical ethic concerning a stipulated freedom for the mastering of nature. It is the entity that has been given a compass to guide us through the difficult questions ahead. Today's Prometheus can be played by the religious community. For those who question the proper relationship between science and religion, Greek mythology reminds us that Epimetheus and Prometheus were brothers. If we are honest with our history, we will realize that religion (particularly Christianity) and science do have familial ties. Regardless of whether or not the scientific community recognizes it, religion *is* Prometheus. It is religion's role to think through the issues beforehand. It is religion's responsibility to develop a new theological framework that will relevantly engage the new world that biotechnology is offering. It is religion's God-given task to offer the world the compass to guide in the decisions that lie ahead.

Becoming Prometheus is the challenge of the religious community, and it has been the effort of this book. In this final chapter, we will examine some of the ways in which the concepts of this book could make an impact in the actual decisions and debates that are ongoing. Before we get to those practical applications, we must stop to recognize, however, that there are obstacles on all sides. Before we consider how this book could be practically applied, we must understand what forces would be working against us. They are the *Obstacles from the Right, Obstacles from the Left,* and *Obstacles from Science itself.*

Obstacles from the Right

An observation of the cultural, religious, and political landscape of the United States would undoubtedly reveal that there is a strong and growing religious conservatism. This movement has tremendous political strength, and seems to be engaged in a battle to preserve what it sees as essentials of life and faith. The battleground is, primarily, the court room and the classroom. From choosing Supreme Court Judges to determining classroom curriculum, this "religious right" movement has been highly active. In many cases, unfortunately, scientific enterprise has been the target of this battle. On one side of a religion-science conflict, there is the battle over our *origins.* In this side of the battle, Creationism and the Intelligent Design theory are seeking to unseat the ruling concepts of

NeoDarwinism. Some books that promote this agenda are <u>Darwin on Trial</u> by Phillip E. Johnson, <u>Darwin's God</u>, by Cornelius G. Hunter, and several books by authors Michael Behe and William Demski.

Not only is there a religious confrontation with science over our origins, but one of equal fervor over our course of future action. In many of the issues that have been discussed in this book, there is a strong voice within the religious right that would see the traffic light (please recall the imagery from chapter 2) as red. For a sample reading of these views, one could turn to <u>Consumer's Guide to a Brave New World</u> by Wesley J. Smith and <u>Life, Liberty and the Defense of Dignity</u>: *The Challenge for Bioethics,* by Leon Kass.

In much of the literature from this religious right perspective, two concepts are often articulated. The first, known by its critics as the "God in the gaps" position, insists that subjects that are beyond some accepted norm of natural knowledge must be the sacred space of God. This is often used to champion the existence of a creator in the gaps of explanation within Darwinism. A second, yet related, concept has been reflected in the term "playing God." In this idea, anything that exists in the perceived gap of natural knowledge is God's sole territory, and should not be encroached by human activity. The mysterious separation between animals and humans, for instance, may present a gap in our natural knowledge. Within the gap, God is inserted. Humans are special because God made us that way. Once God is in the gap, the gap becomes sacred space. Any effort at mixing human and animal genes would be a violation of that sacred space. It would be the unacceptable act of "playing God".

The foundational concepts of this study would not align with this viewpoint. Although issues of biotechnology have been held up to a strong standard of ethics, the "gaps" are not viewed as sacred space. This study, therefore, would embrace many aspects of biotechnology that a significant part of the religious right would not. In order to gain momentum as a true religion-science partnership, the compass designed in this study will have to overcome some significant obstacles from the right.

To those who would oppose this book on those grounds, let us offer this plea: *We have much in common.* We both begin with the fundamental conviction that there is a God, and that God cares about this world and its inhabitance. We both begin with a fundamental conviction that the Bible is truly authoritative, and offers God's Word for human activity and responsibility. We both believe that the material dimension of life interfaces with a spiritual dimension and for that reason, faith and science must engage each other. Given these common points, let us build an atmosphere of trust and dialogue, and be willing to do the hard theological work of interpreting our biblical faith for the emerging world of biotechnology. To address the obstacles from the right, we must begin with the bridges of our commonalities.

Obstacles from the Left

It may be hard to say which came first, but there is no doubt that the efforts of the religious right to dictate ethics and values in government, education, and scientific enterprise is met by strong efforts to keep religion, in any form, out of those areas. All across the United States, there are school boards and government boards trying to draw the line between religion and secular life. As those who represent the religious right fire shots and attacks on both evolutionary science and biotechnology, leaders of these fields respond in kind. In some cases, these responses take the form of scientific apologetics, in which scientists and others write books defending the theories of science. Richard Dawkins is very active on this scene, and has written many books on the subject. Michael Shermer, author of the book How to Debate a Creationist, recently published a new book entitled Why Darwin Matters: Evolution, Design, and the Battle for Science and Religion. The advocacy group Shermer started, known as *The Skeptic Society,* has embarked on a campaign to send a copy of this book to every member of Congress.[196]

In other cases, scientists (and others) choose not so much to win any arguments with religion, but instead argue for a respectful line of separation between the two. In this view, religion should keep out of science not because it is in error, but because its arena of expertise is in a different realm. A chief proponent of this view is the late Stephen Gould, who advocated for the principle of NOMA, which stands for "non-overlapping Magisteria".[197] Some religious leaders also promote this clear separation of the two. The former Bishop of Edinburgh, Richard Holloway, offers such a case in his book Godless Morality. *Keeping Religion out of Ethics*.

At times, however, those who find themselves in the position of defending science against a particular form of religious faith express a generalized loathing of religion in general. This seems best exemplified in the writings of Richard Dawkins. In the introduction, we saw Dawkins calling religion one of the world's greatest evils, and comparing it to the smallpox virus.[198] Dawkins newest book, The God Delusion, pulls no punches from continuing this line of thought. In his opening paragraph following an introduction, Dawkins writes:

> The God of the Old Testament is arguably the most unpleasant character in all of fiction: jealous and proud of it; a petty, unjust, unforgiving control-freak; a vindictive, bloodthirsty ethnic cleanser; a misogynistic, homophobic, racist, infanticidal, genocidal, filicidal, pestilential, megalomaniacal, sadomasochistic, capriciously malevolent bully.[199]

Given the force of Dawkins' apparent loathing of religion, it should not be surprising to learn that Dawkins strongly objects to Gould's respectful truce of the science/religion wars under the rubric of NOMA. Of this idea, Dawkins said that "Gould carried the art of bending over backwards to positively supine lengths."[200] Throughout The God Delusion, Dawkins expends a great deal of energy and his own scientific reputation to disprove the existence of God. It seems, however, that most people agree that the existence of God is beyond scientific analysis, even though the observation of religion as a natural phenomenon is quite possible. Dawkins' attacks on religion have often gone beyond his scope as a scientist, causing some erosion of his reputation. With this recent effort to disprove God, Dawkins may risk losing an audience that has, to this point, respected what he has had to say.

From the left, then, there is a strong effort to keep religion out of any involvement with the progress of scientific inquiry or biotechnology developments. Even though the compass developed in this book does not reflect the anti-science position mentioned earlier, it would still get caught up in the net of this view from the left. Given this obstacle, how could this compass gain appropriate use?

On the one hand, it would be important for us to support the domain of scientific inquiry free from religious constraints. Just as Dawkins is fundamentally in error in his attempt to use science to disprove God; it is an equal error for religion to impose the leap of faith necessary to believe in God upon scientific inquiry. The

Intelligent Design theory, for instance, is an intriguing viewpoint for philosophy and religious studies, but does not belong in the science classroom.

Beyond the support of protecting the realm of scientific inquiry from the constraints of religious imposition, however, there are areas in which we would need to advocate for a respectful, cooperative and constructive engagement between science and ethics. When one moves from the realm of pure "science", which is primarily an activity of acquiring knowledge; to that of biotechnology, which is an application of that knowledge; the scientific enterprise ceases to be in its own environment. In the efforts of application, the enterprise has become a part of the total community of life. At that point, the issues of ethics become a valid qualifier on the pursuits of the enterprise. The compass developed in this book has lifted up two primary ethical qualifiers: love and justice. Granted, we arrived at these qualifiers through a clearly religious perspective. Even if one would discount the path that led us to these values, there still could be a general acceptance of them as being valid measurements of evaluation. In other words, the compass could be employed in a purely secular environment, in which the acceptance of God or biblical authority is not a given. All that would be necessary is an agreement that love and justice, as defined in this book, are valid goals for ethical alignment. In referencing the work of CS Lewis in his book <u>Mere Christianity</u>, Francis S. Collins argues that there are universal values accepted by most people, regardless of religious affiliation, or lack thereof. Collins calls this "the moral law". He writes:

> In the area of medicine, furious debates currently surround the question of whether or not it is acceptable to carry out research on human embryonic stem cells. Some argue that such research violates the sanctity of human life; while others posit that the potential to alleviate human suffering constitutes an ethical mandate to proceed. Notice that in all (sic) these examples, each party attempts to appeal to an unstated higher standard. This standard is the Moral Law.[201]

The existence of a "moral law" as a universal guide does not receive wholesale acceptance. Some argue against its true existence, while others propose explanations for an apparent "moral law" in evolutionary processes. Collins himself recognizes this, even as he posits the concept.[202] Although we may not be able to point to some universally accepted moral law, there could be pockets of consensus that would agree that love and justice, as defined in this book, are good ethical measurements of biotechnical issues. In countering the obstacles from the left, then, the compass could stand on its own, without a necessary link to its theological origins.

Obstacles from Science Itself

One of the most active lines of evolutionary theory development today is in the area of religion. From a variety of directions, religiosity is under the microscope. From one angle, sociobiology and evolutionary psychology experts are proposing theories that demonstrate religion as an aspect of group selection processes, in which religion serves as the collective mind of human communities much like can be observed in the group think of bee colonies or ant mounds. This view is well articulated by David Sloan Wilson in his book Darwin's Cathedral. *Evolution, Religion, and the Nature of Society*. The newest book by Daniel C. Dennett, Breaking the Spell. *Religion as a Natural Phenomenon*, begins with the evolutionary theories but then moves toward an evaluation of the current religious scene.

Whereas some are investigating the origin of religion through evolutionary process theory, others are taking a different course to the same possible conclusion. Hot of the press in this early part of 2007, Ape expert Barbara J King has released a book entitled Evolving God. *A Provocative View on the Origins of Religion.* In studying the brain capacities of different primates in relation to their abilities and complexities of socio-emotional ties, King presents a theory concerning the rise of religious thought in the increasing brain capacity of the human ape as evidenced in cave drawings and archeological finds. Her investigation is intriguing and illuminating.

While some are looking at evolutionary theories and some at ancient cave drawings, still others are looking within the human genome to find clues to the rise of religion. As already seen, Geneticist Dean Hamer has announced the discovery of a genetic link to religiosity in his book The God Gene. *How Faith is Hardwired into our Genes*. Hamer's research has not received universal acceptance, but his book has been widely read.

All of these efforts illustrate that scientific inquiry is on the hunt to explain religion as a natural phenomenon, rising up through natural evolutionary processes, and existing in the physical matter of the human brain and the human genome. Given the strength of these early discoveries, it seems quite possible that religion will, indeed, be proven to be a truly natural phenomenon. This leads some to conclude that religion will be permanently and totally defeated, once its supposed mystical origins have been unmasked. Harvard Professor E.O. Wilson has stated:

> We have come to the crucial stage in the history of biology when religion itself is subject to the explanations of the natural sciences. Sociobiology can account for the very origin of mythology by the principle of natural selection acting on the genetically evolving structure of the human brain. If this interpretation is correct, the final decisive edge enjoyed by science will come from its capacity to explain traditional religion, its chief competitor, as a wholly material phenomenon.[203]

It is this line of thought that would present an obstacle to the utilization of this book and its compass. One could adopt the view that if religion is proven to be a "wholly material phenomenon", then all that religion holds to be true is proven false; Indeed, God is proven false.

In response to the obstacles presented by science itself, the first thing we should do is to embrace good, sound scientific theory concerning the nature of religion, even if it does threaten our traditional views. If this causes a need for us to rethink matters of theology, and particularly religious anthropology, so be it. Yet even as we are willing to make changes within our thinking, there is no need to abandon our belief in a God who cares about what happens in this universe. Religion, as we know it, is but one side of a relationship. It is the way in which human beings conceptualize a God. Science can dissect that side of the relationship, and lay it out for all to see. All that will be proven, however, is that the human side of the God-human relationship has arisen through natural processes. This discovery will not, in any way, speak to the existence of the one who is on the other side of that relationship,

God. In addressing this issue in relationship to the discovery of the so called "God Gene", Hamer writes:

It is essential to realize that there is nothing intrinsically theistic or atheistic about postulating a specific gene and biochemical mechanism for spirituality. If God does exist, he would need a way for us to recognize his presence. Indeed, many religious believers have interpreted the brain-scan experiments as supporting the existence of a deity; why else, they ask, would we have a "God module"prewired in the brain? If, on the other hand, there is no God, then all religious and mystical experiences may represent no more than the random firing of poorly programmed neurons in the brain. Science can tell us whether there are God genes but not whether there is a God.[204]

In response to the obstacles presented by science, we should on the one hand, welcome new discoveries even if they threaten traditional viewpoints, while at the same time, keeping science clear about its own limits at addressing the existence of God, and of God's desire to guide human activity. We will, in fact, need to become more skilled at our apologetics in response to the emergence of theories that present religion as a material and natural phenomenon.

In an effort to promote the use of our compass, we would run into significant resistance from several sides. We would have to navigate through the obstacles presented from the right, from the left, and from science itself. This would be no easy task. Beyond these challenges, however, lies the real charge. How could this compass gain use beyond the pages of this book? How could we become Prometheus?

Toward Strategies of Application

The issues of biotechnology are incredibly important for our time. The *Wall Street Journal* quote posted on the front cover of the President's Council on Bioethics report entitled <u>Beyond Therapy</u> states that "in terms of their importance to the future of our society, these issues rank up there with war and peace."[205]   Given   the importance of these issues, it is imperative that religious communities take on the role of becoming Prometheus for our society. To be the Church in the world today, we must lead the way toward a responsible process of thinking through the ethical questions before the leaps into biotechnical frontiers. With the compass to guide us, we could partner with political leaders and the scientific community in helping to shape the direction that biotechnology will take us. In becoming Prometheus, I would like to offer the following suggestions:

1: Stop focusing on trivial matters, and start addressing the major issues

I was recently invited to offer a series of lectures on the subject of bioethics at a large, wealthy church in the suburbs of Detroit. As the pastor of that church reflected on the weekend at its conclusion, he said: "These are the issues we should be addressing, but all our time is spent on things that seem so trivial in comparison." Indeed, that seems to be a problem in many churches across the country. Today churches are embroiled in internal conflicts over issues such as music styles, generational tensions, and budget applications. With so much time spent in these areas, many churches have no time left over to do the necessary work of holding up the compass for the wider community. In order to become Prometheus, we will need to let go of the internal matters that drain our energies and time. Letting go of these internal issues will free the church to look out into the world, which is precisely where the call to be Prometheus leads.

2: Become educated!

The world of biotechnology is evolving fast, and it can be very difficult to keep up with current discoveries and achievements. In order to become Prometheus, religious leaders and communities will need to engage in an ongoing process of education. Failure to do so will result in an inability to speak relevantly, or could lead to uninformed stances that are emotionally strong but content weak. Recently, for example, I heard a church member criticize another because that person was utilizing stem cell therapy to combat a form of leukemia. She had no idea of the difference between adult stem cells and embryonic stem cells, and was under the impression that the whole subject was ethically mired. For that to happen on an individual basis can certainly be expected; but if a religious community fails to engage in the necessary activity of education in order to enlighten its people and shape its voice to the external community, it is doing a disservice to everyone. Added to the curriculum of every church, I propose, should be courses that educate people on current aspects of biotechnology, and offer people practice in applying the compass to evaluate these issues. Education is paramount.

3: Develop a network of relationships

We should not expect that when the ethics of biotechnology are being discussed in a scientific or political environment, there will always be a seat reserved for representatives of the religious community. To the degree that this does happen, we can give thanks. It seems likely in the future, however, that this entitlement to participate may not be so readily offered. The increasing marginalization of the church in society, the widening gulf between the limited expertise of religious leaders in relationship to the complex issues at hand, and the growing theories of religion as an evolved and wholly material phenomenon, could all contribute to a decrease of entitlement for religious leader participation in discussions of bioethics. Where entitlement wanes, relationships must take over. Religious leaders must proactively seek to establish a network of relationships with leaders of science, medicine, government, and leaders of biotech corporations. Within the framework of relationships, two things will happen. First, the religious leader will become more informed about the real issues of biotechnology. Second, the religious leader will become known, trusted, and even sought-out for discussions of ethical matters.

4: Lead by example

If the religious community wishes to hold up the core values of love and justice as a guiding compass for biotechnology, we must first demonstrate that in all we do, we are seeking to align ourselves with these values. The church should orient its time, energies, and finances toward a selfless attention to the needs of others, particularly the most vulnerable and forgotten. We should seek to work for issues of justice on the local, national, and world-wide scene. In doing this, two things would be accomplished. First, we would perfect the guidance of the compass, and teach it to others not simply by words, but by modeling behavior. Second, an alignment with love and justice would reduce our internal focus on the more trivial matters, and enable us to better enact the first suggestion that has been offered.

5: Train experts in theology and biotechnology

In the reading and research that I have done to prepare this book, I have discovered that there are far more scientists who are addressing theology than there are theologians addressing science. There are, to be sure, many theologians who speak to matters of science, but very few who are trained in matters of science or biotechnology *and* trained in matters of theology. Many signs indicate that the 21st century will be one in which biotechnology will take a large role. It would behoove the religious community to develop tracks of education that equip "biotechnical theologians". Seminaries should add more bioethical courses to their curriculums, and even consider developing a new degree program in bioethics. Such an education could (and I think, *should*) be interdisciplinary. Perhaps seminaries could partner with technology schools and medical schools to provide a track in which a seminarian pursuing a bioethics degree would actually take courses in these other institutions. A cadre of theologian-scientists, or scientific-theologians, would greatly aid the religious community in becoming Prometheus.

Conclusion

These five suggestions are not exhaustive. They are simply offered to get the conversation started. Becoming comfortable about our own ability to evaluate biotechnical issues with the compass of this book is our first challenge. Finding our place in society in order to hold up that compass to others is our second challenge. Both challenges, however, are vitally important as we take on the role of leadership in the ethical challenges of our day. Epimetheus may have been the keeper of the box of godly powers, but his brother Prometheus could have helped him make the right decisions. Science and Faith, working together in mutual respect and appreciation, can enable a bright future in which many dreams of biotechnology are realized, and carried out in wonderful acts of love and justice for the world. Let us build a new relationship with our sibling science. Let us fully embrace the freedom that God has given. Let us hone our use of the compass. Let us go boldly into God's future, *navigating through a stipulated freedom*.

Chapter 7 Notes

[194] http://www.newscientist.com/article.ns?id=dn3217

[195] http://www.cnn.com/2003/HEALTH/01/04/human.cloning/index.html

[196] http://www.skeptic.com/eskeptic/06-02-09.html

[197] Gould, Stephen Jay. Rocks of Ages. *Science and Religion in the Fullness of Life*, p. 49

[198] See page 4.

[199] Dawkins, Richard. The God Delusion, p.31

[200] Ibid, p. 55

[201] Collins, Francis S. The Language of God. *A Scientist Presents Evidence for Belief*, p. 22.

[202] Ibib

[203] Wilson, Edward O. On Human Nature. p. ??

[204] Hamer, Dean H. The God Gene. *How Faith is Hardwired into our Genes.*, p.211

[205] Kass, Leon (Ed). Beyond Therapy. Biotechnology and the Pursuit of Happiness. *A Report by the President's Council on Bioethics.*, front cover.

# Bibliography

Alexander, Denis. Rebuilding the Matrix. *Science and Faith in the 21ˢᵗ Century.* Grand Rapids, MI: Zondervan Press, 2001.

Anderson, Walter Truett. Evolution Isn't What It Used to Be. *The Augmented Animal and the Whole Wired World.* New York: W.H. Freeman and Company, 1996.

Barclay, William. The Ten Commandments. Louisville: Westminster John Knox Press, 1998.

Bennett, W. H. (Ed). The New Century Bible Commentary on Exodus. New York: Oxford University Press, 1900.

Bettenson, Henry (Ed). Documents of the Christian Church. Oxford: Oxford University Press, 1963.

Bodmer, Walter and McKie, Robin. The Book of Man. *The Human Genome Project and the Quest to Discover Our Genetic Heritage.* Oxford: Oxford University Press, 1994

Borgmann, Albert. Holding on to Reality: *The Nature of Information at the Turn on the Millennium.* Chicago: The University of Chicago Press, 1999.

Brooke, John and Cantor, Geoffrey. Reconstructing Nature. *The Engagement of Science and Religion.* Oxford: Oxford University Press, 1998.

Brown, William P. (Ed) The Ten Commandments. *The Reciprocity of Faithfulness.* Louisville, Kentucky: Westminster John Knox Press, 2004.

Brueggemann, Walter. Old Testament Theology. *Essays on Structure, Theme, and Text.* Minneapolis: Fortress Press, 1992.

Brueggemann, Walter. (et al) To Act Justly, Love Tenderly, Walk Humbly: *An Agenda for Ministers.* New York: Paulist Press, 1986.

Buber, Martin. I and Thou. Translated by Walter Kaufmann. New York: Charles Scribner's Sons, 1970.

Chalmers, A.F. What is this Thing Called Science? (Third Edition) Buckingham, UK: Open University Press, 1999

Childs, Brevard. Myth and Reality in the Old Testament. Naperville, Ill: A.R.Allenson, 1960.

Chilton, Bruce and McDonald, J.I.H. Jesus and the Ethics of the Kingdom. Grand Rapids, MI: William B. Eerdmans Publishing Company, 1987

Clines, David J. *Sacred Space, Holy Places, and Suchlike*. In On the Way to the Postmodern: Old Testament Essays 1967-1998, Volume 2. Sheffield: Sheffield Academic Press, 1998.

Cole-Turner, Ronald. The New Genesis. *Theology and the Genetic Revolution*. Louisville: Westminster/John Knox Press, 1993.

Collins, Francis S. The Language of God. *A Scientist Presents Evidence for Belief.* New York: Simon and Schuster Free Press, 2006.

Collins, Raymond F. Christian Morality. *Biblical Foundations*. Notre Dame: University of Notre Dame Press, 1986.

Davidman, Joy. Smoke on the Mountain. *An Interpretation of the Ten Commandments*. Philadelphia: The Westminster Press, 1953

Davies, Paul. God and the New Physics. New York: Simon and Schuster, 1983.

Dawkins, Richard. The Blind Watchmaker. *Why the Evidence of Evolution Reveals a Universe without Design.* New York: Norton and Company, 1996.

Dawkins, Richard. The God Delusion. New York: Houghton Mifflin Company, 2006.

Dawkins, Richard. The Selfish Gene. Oxford, England: Oxford University Press, 1976.

Dembski, William A. and Ruse, Michael. The Stem Cell Controversy: *Debating the Issues*. New York: Prometheus Books, 2003

Dennett, Daniel C. Breaking the Spell. *Religion as a Natural Phenomenon*. New York: The Penguin Group, 2006.

Dover, Gabriel. Dear Mr. Darwin. *Letters on the Evolution of Life and Human Nature*. Los Angeles: University of California Press, 2000.

Driver, S. R. The Book of Exodus. Cambridge: University Press, 1918

Duncan, David Ewing. The Geneticist Who Played Hoops with My DNA...and other Masterminds from the Frontiers of Biotech. New York: HarperCollins Books, 2005

Durham, John I. Word Biblical Commentary Volume 3. Exodus. Waco, Texas: Word Books, 1987.

Dyson, Freeman J. Infinite in All Directions. New York: HarperCollins, 1989

Dyson, Freeman J. The Sun, The Genome, and The Internet. Tools of Scientific Revolutions. Oxford: Oxford University Press, 1999

Eliade, Mircea. The Sacred and the Profane. The Nature of Religion. London: Harcourt, Inc., 1957

Erman, Adolf. Life in Ancient Egypt. Translated from the German by H.M. Tirard, New York: Dover Publications Inc, 1971

Ferngren, Gary B (Ed). Science and Religion. A Historical Introduction. Baltimore: The John Hopkins University Press, 2002.

Ferris, Timothy (Ed) The Best American Science Writing 2001. New York: HarperCollins, 2001

Fincham, J. R. S. and Ravetz, J.R. Genetically Engineered Organisms. Benefits and Risks. Buckingham, UK: Open University Press, 1990.

Fretheim, Terence E. Exodus. Interpretation Commentaries. Louisville, Kentucky: John Knox Press, 1991

Friedberg, Errol C. The Writing Life of James D. Watson. New York: Cold Springs Harbor Laboratory Press, 2005.

Fukuyama, Francis. Our Posthuman Future. Consequences of the Biotechnology Revolution. New York: Picador Press, 2002.

Gilkey, Langdon. Nature, Reality, and the Sacred. The Nexus of Science and Religion. Minneapolis: Fortress Press, 1993.

Goppelt, Leonhard. Theology of the New Testament. Volume 1: The Ministry of Jesus in its Theological Significance. Grand Rapids, MI: William B. Eerdmans Publishing Company, 1981.

Gottwald, Norman K. The Hebrew Bible. A Socio-Literary Introduction. Philadelphia: Fortress Press, 1985.

Gould, Stephen Jay. Rocks of Ages. Science and Religion in the Fullness of Life. New York: Ballantine Books, 1999

Gould, Stephen Jay. Wonderful Life. The Burgess Shale and the Nature of History. New York: Norton and Company, 1989

Greeley, Andrew M. The Sinai Myth. New York: Doubleday & Co., 1972

Green, Jay (Ed) The Interlinear Hebrew/Greek English Bible. Wilmington, Del: Associated Publishers and Authors, 1976

Hall, Stephen S. Merchants of Immortality. *Chasing the Dream of Human Life Extension.* New York: Houghton Mifflin Company, 2003

Hamer, Dean H. The God Gene. *How Faith is Hardwired into our Genes.* New York: Bantam Dell Publishing, 2004.

Harrelson, Walter. The Ten Commandments and Human Rights. Philadelphia: Fortress Press, 1980

Hawkins, Jeff. On Intelligence: *How a New Understanding of the Brain will Lead to the Creation of Truly Intelligent Machines.* New York: Times Books, 2004.

Heidegger, Martin. The Question Concerning Technology *and other Essays.* New York: Harper and Row, 1977.

Hendricks, Obery M. The Politics of Jesus. *Rediscovering the True Revolutionary Nature of Jesus' Teachings and How They Have Been Corrupted.* New York: Doubleday, 2006.

Herzfeld, Noreen L. In Our Image: *Artificial Intelligence and the Human Spirit.* Minneapolis: Fortress Press, 2002.

Holloway, Richard. Godless Morality: *Keeping Religion out of Ethics.* London, Canongate Books, 2000.

Hornung, Erik. Conceptions of God in Ancient Egypt. *The One and the Many.* Translated from the German by John Baines. Ichaca, New York: Cornell University Press, 1982

Hunter, Cornelius G. Darwin's God. *Evolution and the Problem of Evil.* Grand Rapids, MI: Brazos Press, 2001.

Janzen, J. Gerald. Exodus. Louisville, Kentucky: Westminster John Knox Press, 1997

Johnson, Philip E. Darwin on Trial. Downers Grove, IL: Intervarsity Press, 1993.

Kadavil, Mathai. The World as Sacrament. *Sacramentality of Creation from the Perspectives of Leonardo Boff, Alexander Schmemann, and Saint Ephrem.* Leuven, Belgium: Peeters Press, 2005.

Kass, Leon (Ed). Beyond Therapy. Biotechnology and the Pursuit of Happiness. *A Report by the President's Council on Bioethics.* New York: HarperCollins, 2003

Keck, Leander, et al (Editorial Board) The New Interpreter's Bible. Volume 1. Nashville: Abingdon Press, 1994.

Kellert, Stephen R. and Farnham, Timothy J. The Good in Nature and Humanity. *Connecting Science, Religion, and Spirituality with the Natural World.* London: Island Press, 2002.

King, Barbara J. Evolving God. *A Provocative View on the Origins of Religion.* New York: Doubleday Books, 2007.

Kohlenberger, John R. (Ed). The NIV Interlinear Hebrew-English Old Testament. Volume 1/Genesis-Deuteronomy. Grand Rapids: Zondervan Publishing House, 1978.

Kuhns, William. Environmental Man. New York: Harper and Row, 1969

Kuhse, Helga and Singer, Peter.(Ed.) A Companion to Bioethics. Oxford: Blackwell Publishers Ltd, 2001.

Kuntz, Paul Grimley. The Ten Commandments in History. *Mosiac Paradigms for a Well-Ordered Society.* Grand Rapids, MI: William B. Eerdmans Publishing Co., 2004

Kurzweil, Ray. Fantastic Voyage. *Live Long Enough to Live Forever.* Private Publication of Ray Kurzweil, 2004.

Kurzweil, Ray. The Age of Spiritual Machines. *When Computers Exceed Human Intelligence* New York: The Penguin Group, 1999

Lappe, Marc. Evolutionary Medicine. *Rethinking the Origins of Disease*. San Francisco: Sierra Books, 1994

Lehmann, Paul L. The Decalogue and a Human Future. *The Meaning of the Commandments for Making and Keeping Human Life Human.* Grand Rapids, MI: William B. Eerdmans Publishing Co., 1995.

Lewis, C.S. The Abolition of Man. *How Education Develops Man's Sense of Morality.* New York: MacMillan Publishing Co., 1955.

Lewis, C.S. The Four Loves. London: Geoffrey Bles Press, 1960.

Lochman, Jan Milic. Signposts to Freedom. *The Ten Commandments and Christian Ethics.* Minneapolis: Augsburg Publishing House, 1982

Lockshin, Martin I. (Editor and Translator from the Hebrew) Rashbam's Commentary on Exodus. Atlanta: Scholars Press, 1997

Lossky, Vladimir. The Mystical Theology of the Eastern Church. London: James Clarke and Co., LTD. 1944

Lovett, William (ED) <u>Martin Heidegger: The Question Concerning Technology</u>. *And Other Essays.* New York: Harper and Row, 1977

Lunbhom, Jack R. *God's Use of the Idem per Idem to Terminate Debate.* In the <u>Harvard Theological Review</u>, vol. 71. 1978

Naisbitt, John and Patricia, Aburdene. <u>Megatrends 2000.</u> New York: William Morrow and Co., 1990.

The National Research Council and Institute of Medicine of the National Academies. <u>Guidelines For Human Embryonic Stem Cell Research</u>. Washington, DC: The National Academies Press, 2005.

Neuhaus, Richard J. <u>Doing Well and Doing Good.</u> *The Challenge to the Christian Capitalist.* New York: Doubleday, 1992

Nielsen, Eduard. <u>The Ten Commandments in New Perspective</u>. London: SCM Press LTD, 1968

Metzner, Ralph. <u>Green Psychology.</u> *Transforming Our Relationship to the Earth.* Rochester, Vermont: Park Street Press, 1999

McCabe, Herbert. <u>God Still Matters</u>. London: Continuum, 2002.

MacIntyre, Alasdair. <u>After Virtue.</u> *A Study in Moral Theory.* (second edition). Notre Dame: University of Notre Dame Press, 1984

McFague, Sallie M. <u>Super Natural Christians.</u> *How We Should Love Nature*. Minneapolis: Fortress Press, 1997.

McNeill, John T (Ed). <u>Calvin: Institutes of the Christian Religion.</u> Philadelphia: The Westminster Press, 1960.

Moehlman, Conrad Henry. <u>The Story of the Ten Commandments.</u> *A Study of the Hebrew Decalogue in its Ancient and Modern Application*. New York: Harcourt, Brace and Co., 1928.

Moltman, Jurgen. <u>The Church in the Power of the Spirit.</u> *A Contribution to Messianic Ecclesiology.* San Francisco: Harper Collins, 1977.

Murphy, Nancey and Ellis, F.R. George. <u>On the Moral Nature of the Universe.</u> *Theology, Cosmology, and Ethics.* Minneapolis, Fortress Press, 1996.

Nelson-Pallmeyer, Jack. <u>Jesus Against Christianity.</u> *Reclaiming the Missing Jesus* Harrisburg, PA: Trinity Press International, 2001.

O'Meara, Thomas F. and Weisser, Donald M. (ED) <u>Projections: Shaping an American Theology for the Future.</u> New York: Doubleday, 1970

Parsons, Ann B. <u>The Proteus Effect:</u> *Stem Cells and their Promise for Medicine.* Washington, DC: Joseph Henry Press, 2004

Peacocke, Arthur. Theology for a Scientific Age. *Being and Becoming-Natural, Divine, and Human.* Minneapolis: Fortress Press, 1993.

Peters, Karl E. Dancing with the Sacred. *Evolution, Ecology and God.* Harrisburg, PA: Trinity Press International, 2002.

Peters, Ted. (Ed). Genetics. *Issues of Social Justice.* Cleveland, Ohio: The Pilgrim Press, 1998.

Peters, Ted. (Ed). Playing God? *Genetic Determinism and Human Freedom.* New York: Routledge Press, 1997.

Philippou, A.J. (ED) The Orthodox Ethos: Essays in Honour of the Centenary of the Greek Orthodox Archdiocese of North and South America Oxford: Holywell Press, 1964.

Propp, William C. The Anchor Bible Commentary on Exodus 1-18. New York: Doubleday Press, 1999

Ratzsch, Del. Nature, Design, and Science. The Status of Design in Natural Science. Albany, NY: State University of New York Press, 2001.

Ratzsch, Del. Science and Its Limits. *The Natural Sciences in Christian Perspective.* (Second Edition). Leicester, England: Apollos Press, 2000.

Rawls, John. Justice and Fairness. Cambridge, MA: The Belknap Press of Harvard University Press, 2001

Reichenbauch, Bruce R. and Anderson, V. Elving. On Behalf of God. *A Christian Ethic for Biology.* Grand Rapids: William B. Eerdmans Publishing Co., 1995.

Reiss, Michael J. and Straughan, Roger. Improving Nature? *The Science and Ethics of Genetic Engineering.* Cambridge, UK: Cambridge University Press, 1996.

Richardson, W. Mark et al (Editors) Science and the Spiritual Quest. *New Essays by Leading Scientists.* London: Routledge Press, 2002.

Ridley, Matt. The Agile Gene: *How Nature Turns on Nurture.* New York: HarperCollins, 2003.

Ridley, Matt. Genome: *The Autobiography of a Species in 23 Chapters.* New York: HarperCollins, 1999.

Robertson, D.B. (Ed). Love and Justice. *Selections from the Shorter Writings of Reinhold Niebhur.* Louisville, Kentucky: Westminster John Knox Press, 1957

Rowland, Wade. Galileo's Mistake. *A New Look at the Epic Confrontation between Galileo and the Church*. New York: Arcade Publishing, 2001.

Sachs, John R. The Christian Vision of Humanity. *Basic Christian Anthropology*. Collegeville, Minnesota: The Liturgical Press, 1991.

Sarna, Nahum M. The JPS Torah Commentary: Exodus. Jerusalem: The Jewish Publication Society, 1991

Schaeffer, Francis A. Pollution and the Death of Man. *The Christian View of Ecology.* Wheaton, Illinois: Tyndale House Publishers, 1970.

Schnackenburg, Rudolf. The Moral Teaching of the New Testament. London: Burns & Oates Publishers, 1964.

Schroeder, Gerald L. The Science of God. *The Convergence of Scientific and Biblical Wisdom.* New York: Broadway Books, 1997.

Smith, Huston. Why Religion Matters. *The Fate of the Human Spirit in an Age of Disbelief.* San Francisco: HarperCollins, 2001.

Sober, Elliot and Wilson, David Sloan. Unto Others. *The Evolution and Psychology of Unselfish Behavior.* Cambridge, Mass: Harvard University Press, 1998.

Stamm, Johann Jakob. The Ten Commandments in Recent Research. Translated from the German by Maurice Edward Andrew. London: SCM Press LTD, 1967

Tillich, Paul. Love, Power and Justice. Oxford: Oxford University Press, 1954.

Vawter, Bruce. On Genesis: A New Reading. New York: Doubleday & Co., 1977

Von Rad, Gerhard. Old Testament Theology. Vol. 1. *The Theology of Israel's Historical Tradition.* Louisville: Westminster John Knox Press, 2001.

Wallace, Ronald S. The Ten Commandments. *A Study of Ethical Freedom.* Grand Rapids, MI: WM B. Eerdmans Publishing Co., 1965

Watson, James D. DNA. *The Secret of Life.* New York: Alfred A. Knopf, 2003.

Wertheim, Margaret. The Pearly Gates of Cyberspace. *A History of Space from Dante to the Internet.* London: W.W. Norton and Company, 2000.

Whitehouse, W.A. Creation, Science and Theology. *Essays in Response to Karl Barth.* Grand Rapids, Michigan: William B. Eerdmans Publishing Company, 1981.

Wickler, Wolfgang. The Biology of the Ten Commandments. Translated from the German by David Smith. New York: McGraw-Hill Book Co., 1972

Wilson, David Sloan. Darwin's Cathedral. *Evolution, Religion, and the Nature of Society.* Chicago: The University of Chicago Press, 2002.

Wilson, Edward O. Consilience. *The Unity of Knowledge.* New York: Vintage Books, 1998.

Wilson, Edward O. The Diversity of Life. Cambridge: Harvard University Press, 1992

Wilson, Edward O. On Human Nature. Cambridge: Harvard University Press, 1978.

Wingerson, Lois. Unnatural Selection. *The Promise and the Power of Human Gene Research.* New York: Bantam Books, 1998

Index

Agape, 86, 87, 88, 91, 92, 100, 103, 104, 109, 112, 150, 154, 159, 161
ancient repetitive elements (AREs), 110
anthropology, 3, 24, 26, 27, 145, 147, 162, 174
artificial intelligence, viii, 7, 147, 151, 152, 153, 154, 156, 161, 163, 166

bionics, viii, 7, 142, 143, 145, 147, 148, 149, 150, 151, 156
biosteel, 107
biotechnology, vii, viii, ix, 1, 2, 3, 4, 5, 6, 7, 8, 9, 10, 11, 14, 16, 17, 18, 19, 22, 26, 27, 33, 34, 35, 36, 39, 40, 48, 49, 59, 64, 65, 66, 71, 74, 77, 78, 79, 88, 91, 92, 93, 94, 97, 98, 99, 100, 101, 103, 104, 105, 106, 108, 111, 114, 117, 132, 135, 138, 141, 145, 149, 150, 162, 163, 165, 166, 167, 169, 170, 171, 172, 175, 176, 177, 178
Boff, Leonardo, 19, 20, 21, 28, 30, 37, 38, 184
Buber, Martin, 17, 23, 37, 109, 181
burning bush, 9, 10, 35, 40, 41, 46, 47, 48
Bush, George W., 58, 59, 89, 103, 127, 135

Christian theology, 1, 3, 4, 15, 56, 81, 145, 160
cloning, 1, 23, 59, 61, 64, 105, 106, 107, 109, 114, 115, 119, 120, 124, 126, 179
consubstantiation, 15
cybernetic, viii, 21, 146, 148, 156, 157, 158, 159, 160, 161, 162
cybernetic immortality, 156
cyborg, 146, 147, 149, 150, 163

Darwin, Charles, 2, 24, 37, 169, 170, 173, 182, 184, 189
Darwinism, 22, 24, 37, 169
Dawkins, Richard, 2, 154, 163, 170, 171, 179, 182
Decalogue, vii, 49, 50, 51, 52, 54, 55, 56, 57, 58, 59, 60, 65, 66, 68, 77, 80, 185, 186
divine value, 10, 12, 13, 15, 18
DNA, 11, 16, 24, 33, 34, 40, 58, 65, 105, 110, 125, 126, 129, 130, 131, 182, 188
Dolly, 105, 120, 124

Egypt, 41, 42, 43, 44, 48, 51, 54, 67, 90, 183, 184
Eliade, Mircea, 13, 27, 28, 36, 37, 183
embryo, human, 59, 117, 119, 120, 121, 124, 126, 127, 138
embryonic cell, 10, 16, 118, 119, 121, 124, 125, 126, 127, 129, 172, 176

freedom, stipulated, viii, 4, 35, 36, 40, 41, 47, 48, 61, 66, 70, 77, 78, 79, 80, 83, 86, 88, 90, 92, 93, 94, 98, 100, 101, 103, 104, 108, 109, 114, 117, 121, 126, 131, 132, 136, 138, 145, 147, 149, 150, 159, 168, 178

genetic modification, 6, 97, 99, 103, 106, 107, 112
genetically modified
    animals, 107, 109
    corn, 100
    crops, 101
    plants, 98, 99, 100, 101, 102, 103
    wheat, 97
GM. See genetically modified

God's core values, vii, viii, 71, 76,
91, 92, 93, 102, 103, 105, 107,
108, 109, 111, 112, 122, 123, 134,
137, 145, 148, 150, 153, 155, 159,
160, 163
Greek mythology, 6, 7, 49, 95, 117,
129, 141, 163, 165, 166, 167, 168

Hades, 6, 7, 138, 141, 145, 158, 161
Hamer, Dean, 24, 37, 173, 175, 179,
184
Harrelson, Walter, 54, 61, 63, 67, 68,
184
Hebrew Scriptures, 7, 9, 12, 13, 14,
15, 44, 45, 47, 57, 60, 71, 72, 73,
74, 79, 83, 84
Heidegger, Martin, 19, 20, 23, 37,
110, 184, 185
Hendricks, Obrey M., 71, 72, 73, 76,
84, 95, 184
Hephaestus, 6, 7, 138, 141, 151
Hercules, 6, 114, 117, 129
Hermes, 6, 114, 117, 129
hierophany, 13
Hittite treaty, 52, 53
humanzee, 79, 92, 108

idem per idem, 46, 47
in vitro fertilization, 119, 120, 121
intrinsic, vii, 10, 12, 13, 15, 16, 17,
18, 19, 31, 32, 33, 34, 38, 64, 65,
77, 81, 127, 128, 175

Jesus Christ, 30, 55, 56, 71, 72, 73,
79, 80, 81, 83, 84, 85, 86, 95, 96,
182, 183, 184, 186
Lewis, C.S., 26, 37, 87, 88, 96, 172,
185
Lochman, Jan Milic, 50, 54, 59, 67,
68, 185

Mishpat, 73
mysterion, 30, 31

Nelson-Pallmeyer, Jack, 71, 72, 73,
95, 186

New Testament, 14, 15, 29, 55, 70,
79, 80, 81, 82, 83, 84, 85, 86, 87,
95, 96, 183, 188

Peacocke, Arthur, 18, 20, 37, 186
Peters, Ted, 11, 33, 34, 38, 60, 164,
186, 187
postmodernism, 21, 22, 27
prenatal engineering, 132, 135, 136
Prometheus, 7, 165, 167, 168, 175,
176, 178, 182

religion, 1, 2, 3, 13, 42, 117, 138,
165, 168, 169, 170, 171, 173, 174,
175, 177
robosexuals, 153

Schmemann, Alexander, 30, 37, 38,
184
science, viii, 1, 2, 3, 4, 19, 20, 21, 24,
34, 40, 58, 99, 101, 105, 110, 112,
117, 125, 133, 138, 143, 145, 150,
165, 167, 168, 169, 170, 171, 172,
174, 175, 177, 178
stem cell, vii, 1, 6, 10, 16, 25, 64, 82,
106, 117, 118, 119, 120, 121, 122,
123, 124, 125, 126, 127, 129, 138,
139, 141, 166, 172, 176
multipotent, 119
pluripotent, 119, 122, 124, 126,
127
totipotent, 118

Ten Commandments, 4, 48, 49, 50,
51, 52, 54, 55, 56, 60, 61, 67, 68,
181, 182, 184, 185, 186, 188
theocentric cosmology, 19
transubstantiation, 15

Von Rad, Gerhard, 51, 52, 53, 54, 55,
59, 67, 68, 81, 188

world as sacrament, 30, 31, 39, 65,
129

xenotransplantation, 106

192